Non-Soil Fumigation

A Pesticide Applicator Study Manual

Published by the California Department of Pesticide Regulation

July 2023

Photo and Graphics Credits

Cover Graphics:
Provided by the California Department of Pesticide Regulation

Interior Graphics:
Photograph and graphics credits are noted with the images throughout the manual and in the references. Document design and layout provided by Steve Tomasko.

Adapted by the California Department of Pesticide Regulation. This study manual was developed using content from the Non-Soil Fumigation – A National Pesticide Applicator Study Manual published by the Pesticide Educational Resources Collaborative (PERC)* and includes California specific information relevant to commercial pesticide applicators who intend to use non-soil fumigant products in California.

*This publication was developed under Cooperative Agreement No. X8-82616301, awarded by the U.S. Environmental Protection Agency to the University of California Davis Extension in cooperation with Oregon State University. EPA made comments and suggestions on the document intended to improve its scientific analysis and technical accuracy. However, the views expressed in this document are those of its authors and do not necessarily represent the views and policies of the EPA. EPA does not endorse any products or commercial services mentioned in this publication.

 This work is licensed under a Creative Commons Attribution NonCommercial-ShareAlike 4.0 International License. © 2021 The Regents of the University of California, Davis campus. For information contact PERCsupport@ucdavis.edu.

Contributors

California Department of Pesticide Regulation: Nathan Desjarlais, Enforcement Headquarters Branch; Jessica Teague, Enforcement Headquarters Branch; Peggy Byerly, Enforcement Headquarters Branch; Nino Yanga, Pesticide Programs Division; Mara Johnson, Enforcement Headquarters Branch; Alicia Scott, Enforcement Headquarters Branch, Amanda Gregory, Enforcement Headquarters Branch.

Pesticide Educational Resources Collaborative: Director, Suzanne Forsyth, University of California Davis Extension; Deputy Director, Kaci Buhl, Oregon State University; Project Officers, Matt Lloyd & Stephanie Burkhardt, U.S. EPA, Office of Pesticide Programs.

Primary Authors: Project Lead, Sandra McDonald, Mountain West Pesticide Education & Safety Training; Lead Writer, Steve Tomasko, University of Wisconsin-Madison.

Acknowledgements

This manual was prepared by Sandra McDonald and Steve Tomasko with support provided by the Pesticide Educational Resources Collaborative (PERC) and the United States Environmental Protection Agency (EPA) Office of Pesticide Programs.

Significant credit is given to many state pesticide safety education programs for use of information in their state pesticide safety training manuals, which were used to develop the framework for this study guide. Those programs include:

The University of Arkansas, Division of Agriculture Research and Extension; Ohio Department. of Agriculture; The University of Wisconsin-Madison; Texas A&M Agrilife Extension; Purdue University Cooperative Extension Service; University of Florida Institute of Food and Agricultural Sciences; Montana State University Extension; Iowa State University Extension and Outreach; Nevada State Department of Agriculture; The United States Department of Agriculture.

The following people generously gave their time and contributed their expertise by serving as editors and reviewers:

Alan Brown, ABC Home and Commercial Services
Allison Cuellar, Texas Department of Agriculture
Alison Marwitz, Ecolab
Allen Fugler, HIIG Xterminator Pro
Amanda Bachmann, SDSU Extension
Anne Bookout
Barb Nead-Nylander, Douglas Products
Betsy Buffington, Iowa State University
Bonnie Rabe, Rollins
Bret Allen, Nevada Department of Agriculture
Caroline Kirby, Plunkett's Pest Control
Christina Zimmerman, Washington State Department of Agriculture
Cory Goeltzenleuchter, McCall Service, Inc.
Dan Wixted, Cornell University
Derek Johnson, Gregory Pest Solutions
Derrick Lastinger, Georgia Department of Agriculture
Dale Dubberly, Ensystex
Ed Crow, Pennsylvania State University
Erin Smilanich, State of Minnesota Department of Agriculture
Garnet Cooke, State of Oregon–Oregon OSHA
Jeremy Jackson, UPL
Jim Fredricks, National Pest Management Association
Leslie Talpasanu, California Department of Pesticide Regulation
Linda Johns, University of Minnesota
Michael Murray, Wisconsin Department of Agriculture, Trade and Consumer Protection
Mickey Taylor, The University of Georgia
Neil Richmond, Florida Department of Agriculture and Consumer Services
Patricia Hottel, McCloud Services
Shawn Wilson, Cardinal Products

Table of Contents

Chapter 1: Fumigant Basics — 1
   What Is a Fumigant? — 2
   Physical and Chemical Characteristics of Fumigants — 4
   Movement of Fumigants Through the Application Site — 7
   Commonly Used Non-Soil Fumigants — 8
   Pest Factors That Influence Fumigation — 14

Chapter 2: Fumigant Labels and Fumigant Handling — 19
   California Pesticide Use Requirements — 19
   Fumigation of Enclosed Areas — 20
   Labels and Labeling — 21
   Information on Labels — 21

Chapter 3: Planning for a Fumigation — 29
   Premise Inspection — 29
   Calculating Fumigant Dosage — 32
   Calculating Area and Volume — 32
   Sealing Fumigation Sites — 38
   Fumigant Monitoring — 40
   Notification Requirements — 43

Chapter 4: Fumigation Safety — 45
   Fumigation Safety — 46
   Toxicity, Risk, and Exposure — 46
   How People are Exposed to Fumigants — 46
   Signs and Symptoms of Fumigant Exposure in People — 47
   Steps to Minimize Fumigant Health Effects — 49
   Air Monitoring — 50
   Sensory Irritation — 51
   Secure the Fumigation Site — 52
   Fumigation Management Plans — 54
   Post-Application Summary — 55

Chapter 5: Personal Protective Equipment — 59
   Personal Protective Equipment — 59
   Respirators — 63

Chapter 6: Fumigation Methods and Application Equipment — 73
   Steps to Fumigation — 73
   Fumigation Methods — 73
   Aeration — 78
   Equipment Compatibility and Equipment Failure Concerns — 80

Chapter 7.1: Chamber Fumigation — 83
   Fumigation Chambers — 83

Table of Contents

Chapter 7.2: Commodity, Post-Harvest, and Quarantine Fumigation — 87
- Commodity and Post-Harvest Fumigations — 88
- Quarantine Fumigation — 91

Chapter 7.3: Burrow Fumigation for Vertebrates — 95
- Burrow Fumigation — 95
- Conducting a Burrow Fumigation — 96
- Label Restrictions — 98

Chapter 7.4: Fumigation of Certain Structures — 103
- Building Fumigation — 104

Chapter 7.5: Spot Fumigation — 107
- Spot Fumigation — 107

Chapter 7.6: Fumigation of Transport Vehicles, Containers, and Ships — 111

Chapter 7.7: Sewer Line Root Control — 115
- Purpose of Sewer Line Root Control — 115
- Methods to Control Roots — 118
- Metam Sodium — 119

Chapter 7.8: Sanitizing Wooden Wine Barrels and Wine Bottle Corks — 135
- Sanitizing Wine Barrels and Corks — 135

Chapter 7.9: Remedial Wood Protection — 139
- Remedial Protection of Wood with Fumigants — 139

Chapter 8.1: Termites — 147
- Termites — 147
- Fumigation For Control of Termites — 150

Chapter 8.2: Wood-Boring Beetles — 153
- Wood-Boring Beetles — 153

Chapter 8.3: Stored Commodity Pests — 157
- Pests of Stored Products — 157
- Control of Grain and Commodity Pests — 162

Chapter 8.4: Miscellaneous Pests — 165
- Miscellaneous Pests — 165
- Rodents — 165
- Microorganisms — 168

Correct Answers to Chapter Review Questions — 175
Glossary — G–1
Index — In-1

Introduction

This study manual will prepare individuals for the Department of Pesticide Regulation's (DPR's) commercial applicator Non-Soil Fumigation category (Category M) examination.

DPR's Non-Soil Fumigation category is intended for individuals who perform pest control using a pesticide labeled as a fumigant. This category does not include structural pest control required to be licensed under Chapter 14 (commencing with Section 8500) of Division 3 of the Business and Professions Code.

Individuals who apply to hold or currently hold a DPR-issued Qualified Applicator Certificate (QAC) or Qualified Applicator License (QAL) obtain the Non-Soil Fumigation category through DPR's examination process. Note that each designated qualified applicator for a main business location or branch location must have a QAL.

DPR has established knowledge expectations (KEs) to prepare an individual for taking the Non-Soil Fumigation category examination. The complete list of KEs can be found on DPR's Licensing and Certification website. These KEs were developed in conjunction with this manual and should be used as a supplemental study resource.

This study manual covers the basic concepts of non-soil fumigants including, but not limited to, the following topics: use, handling, active ingredients, planning and application, safety, personal protective equipment, application methods, application equipment, application settings/sites, and pests. DPR does not endorse or recommend the use of any pesticide product or pesticidal device mentioned in this study manual but instead provide them as examples.

In addition to this study manual, we recommend that examinees review DPR's Pesticide Safety Information Series (PSIS) leaflets on Storing, Moving, and Disposing of Pesticides in Agricultural Settings (A-2) or Storing, Moving, and Disposing of Pesticides in Non Agricultural Settings (N-2). These summarize key concepts on the storage, transport, and disposal of pesticides. These PSIS leaflets are available in English, Spanish, Punjabi, and Hmong on DPR's website, and from your local County Agricultural Commissioner's (CAC) office.

This study manual covers important aspects of pesticides used for non-soil fumigations. However, the study manual is not a substitute for reading and understanding the label of the specific fumigant product you will be using. Thoroughly read the label prior to purchase and application. Be aware that pesticide manufacturers may update or change the contents of (or information on or instructions on) a pesticide label to address hazards to handlers, nearby workers, other bystanders, and the environment. Reading the label will inform you of any application restrictions to avoid potential human exposure incidents, protect endangered or non-target species, and to avoid harmful environmental effects from the use of the pesticide. Any use in conflict with labeling is a violation of both California and federal law.

Each chapter ends with review questions. After reading each chapter, test your understanding of the information presented in the chapter by answering the review questions. Correct answers are given on page 175. These review questions are in the same multiple-choice selection format as the questions on the DPR examination. However, these questions are not necessarily examination questions.

CHAPTER 1

Fumigant Basics

LEARNING OBJECTIVES

- ☑ Explain what a fumigant is and specify what makes fumigants different from other pesticides.
- ☑ Explain how fumigants change from a liquid or a solid into a gas.
- ☑ Outline the chemical characteristics of fumigants.
- ☑ Describe and explain the factors that affect movement of fumigants through an application site.
- ☑ List the common fumigants used for non-soil fumigation.
- ☑ Describe the characteristics of common fumigants used for non-soil fumigants.
- ☑ Discuss the importance of accurate pest identification.
- ☑ Explain pest factors that determine use of fumigation as a control method.
- ☑ Explain the importance of choosing the proper application rate and timing of application.

Terms to Know

The following are important terms to know from this chapter. They are explained and *italicized* in the text and defined in the glossary at the end of this manual.

Absorb / Absorption	Desorption	Molecular Weight	Toxic
Aerate / Aeration	Diffuse / Diffusion	Particulate	Vapor
Aerosol	Flammability	Pesticide	Vapor Pressure
Boiling Point	Fumigant	Solubility	Volatile
Chemical Reactivity	Fumigation	Sorption	Volatility
Concentration	Gas / Gases	Specific Gravity	Volatilization
Corrosive	Molecules	Stratify	Warning Agent

This study manual deals with non-soil fumigations, primarily of spaces and stored commodities. These types of applications include fumigations of commodities (food and non-food) in a structure (such as fumigation chambers, shipping containers, or grain silos) or vehicle (such as a boxcar or ship), or under a tarp (inside an enclosed space or outdoors). It also includes fumigations of other spaces or items such as wine barrels, utility poles, planting mediums (i.e., potting soil/mix), rodent burrows, and sewer pipes. In California, the term "soil fumigation" is reserved for applications where soil is treated with a fumigant (e.g., fields, forests, golf courses, individual tree or vine holes, etc.). Applications of fumigants in rodent burrows, sewer pipes that are in soil, and of planting medium, a commodity sometimes called "potting soil,", are considered non-soil fumigation. Many of the fumigants used in soil and non-soil fumigation are similar or the same. However, this study manual focuses only on non-soil fumigation procedures.

What is a Pesticide?

Many people mistakenly use the words "insecticide" and "pesticide" interchangeably. However, the word "pesticide" is a broad term that covers many substances. Insecticides are just one specific type of pesticide. The "cide" part of the word comes from Latin meaning "to kill." Therefore, a pesticide kills pests—a broad category because those pests could be insects, weeds, fungi, mice, or other organisms we consider pests. If we are trying to control weeds, we would use an herbicide. For fungi, we use a fungicide, insecticides are aimed at insects and rodenticides at rodents. These are just four examples of the many types of pesticides in use today.

In California, the Structural Pest Control Board issues the license for the use of non-soil fumigants, such as sulfuryl fluoride, to control structural pests (e.g., termites, powderpost beetles, bedbugs, cockroaches, rodents, etc.) to protect structures. This study guide does not cover these uses.

What Is a Fumigant?

Before we discuss *fumigants*, we need to define the broader term—*pesticide*. A pesticide is any substance used to directly control pest populations or to prevent or reduce pest damage. Not all pesticides kill the target pest—some may only inhibit the growth of the pest, repel it, or reduce its capacity to reproduce.

Fumigants are a type of pesticide with unique physical and chemical characteristics. Fumigants are gases, or turn into gases, after application. Fumigants may be odorless and usually cannot be seen. If you are familiar with non-fumigant pesticides but have not used fumigants, it is very important to understand the distinctive qualities of fumigants because their unique characteristics play a large role in their safe and effective use.

> **Volatile and Words Related to It**
> **Volatile**: A substance that vaporizes readily. That means it changes from a liquid (or even a solid) into the gas phase.
> **Volatility**: A measure of how easily a substance vaporizes.
> **Volatilization**: The process or act of vaporizing.

Fumigants are Gases

Fumigants, by their chemical nature, are *volatile*, meaning they readily turn into a gas (sometimes the words "vapor" and "gas" are used interchangeably. It is this characteristic that makes fumigants so useful in certain situations. As a gas, fumigants are *toxic* when *absorbed* or inhaled.

The above description of fumigants excludes *aerosols*, which are *particulate* suspensions of liquids or solids dis-

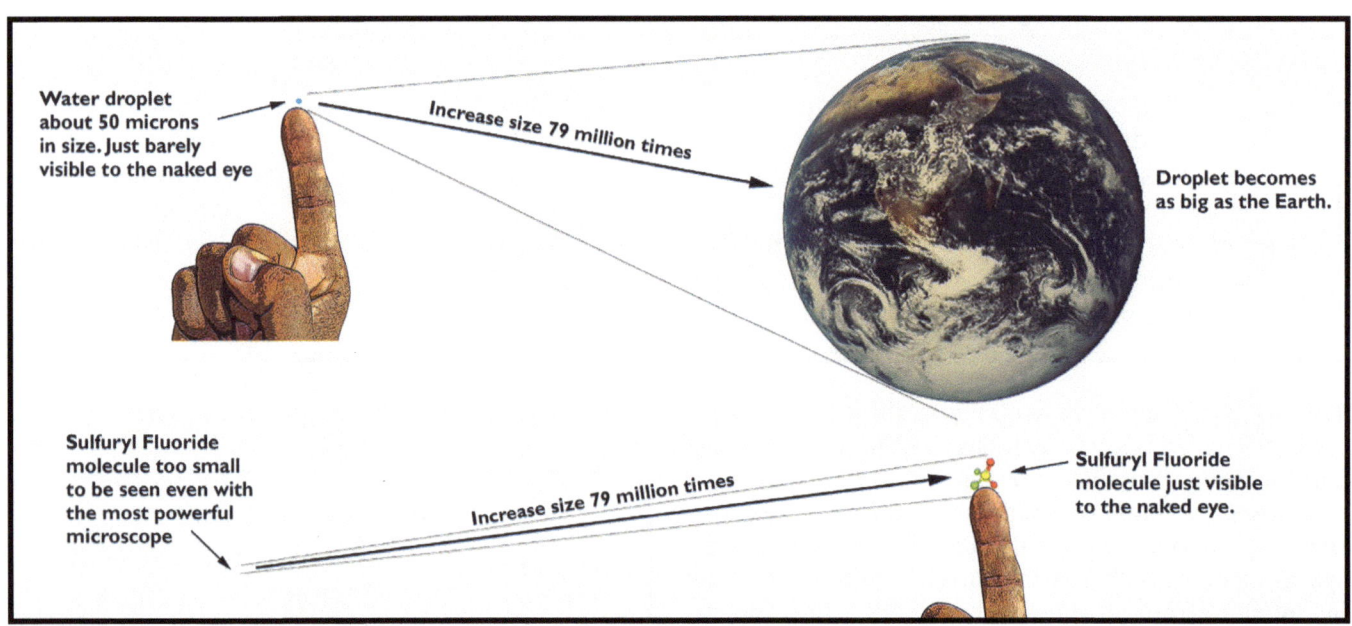

Figure 1: Comparing numbers and size. Extremely large or small numbers and sizes are often hard to comprehend. The graphic above attempts to illustrate how small individual molecules can be (in this case, of the fumigant sulfuryl fluoride). If we take a very small spray droplet of 50 microns in size (smaller than a period at the end of this sentence) and enlarge it 79 million times, it would be about the size of the earth. If we did the same with a molecule of sulfuryl fluoride, it would be barely visible to the naked eye. This hopefully gives an idea of how small fumigant molecules can be, even when compared to a water droplet that we would consider to be a "mist." (Illustration © UW-Madison Pesticide Applicator Training Program).

persed in air (not gasses) that are often referred to as smokes, fogs, or mists. It is important to make this distinction because it emphasizes one of the most unique properties of fumigants—as gases, they spread out through the space in which they are applied as separate *molecules*.

This brings us to a concept that may be difficult to comprehend or visualize—the very large and the very small. Even though a very fine particulate mist droplet or smoke particle seems quite small, they are still massively larger than the individual gas molecules of a fumigant (Figure 1).

Although fumigants are gases when in use, the form you purchase them in varies. Some may come packaged as:
- **Gases:** Carbon Dioxide (CO_2) and sulfur dioxide (SO_2) are kept under pressure as a gas in cylinders. CO_2 is used to treat commodities in storage and containment structures and for outdoor burrowing pest control. SO_2 can be used to fumigate wine barrels and corks, and certain commodities.
- **Liquefied gases:** These are kept under pressure as a liquid in cylinders. They change from a liquid to a gas when released. Methyl bromide and sulfuryl fluoride are typically packaged this way. Some phosphine products, and mixtures of phosphine and CO_2, also come in pressurized cylinders.
- **Volatile liquids:** Some fumigants, such as metam sodium, are liquid at normal temperatures. After exposure to water, metam sodium breaks down to release methyl isothiocyanate (MITC).
- **Solids:** Some fumigants, such as dazomet and aluminum or magnesium phosphide, are packaged as solid pellets or tablets. These react with moisture to release MITC or phosphine gas, respectively.

How Fumigants Kill

Fumigants are a non-selective type of pesticide and kill a wide variety of organisms we might consider pests: insects, plants, nematodes, rodents, and microbes. They also can harm or kill non-target organisms (including people) when not used correctly. Some fumigants control pests by interfering with an organism's respiration (breathing). Other fumigants enter tissues and disrupt the metabolism or other essential processes of animal or plant cells.

The killing action of a fumigant is influenced by its *concentration* in the atmosphere, the length of time it stays in the air, the temperature and humidity of the area, and other factors at the time of fumigation. Because of their nature as a gas, fumigants do not stick around once the fumigation procedure is complete. Therefore, unlike some other types of pesticides, they have no residual property and do not provide any post-application protection from pests.

General Advantages and Disadvantages to Fumigants

As an applicator, you need to clearly understand the potential hazards and problems associated with the use of fumigants. Most fumigants are highly toxic to all forms of life, including humans. Fumigation is a highly technical operation that requires equipment, techniques, and skills not usually associated with other types of pesticides.

Advantages of using fumigants over other pesticides include:
- Fumigants are generally toxic to all pests, such as insects, spiders, mites, rodents, and fungi, specifically, those that are covered in this study manual.
- Fumigants will usually reach pests where you cannot apply sprays, powders, or dusts.
- Some fumigants can be used to kill pests in or near food, leaving no harmful residues while other types of pesticides might contaminate or taint food.
- For certain commodities, fumigation is the only practical way to control pests.

Some disadvantages of fumigants:
- Fumigants are very toxic to people and other non-target organisms.
- Fumigants have no residual pest control action. Once the application site is cleared of fumigant, control ceases.
- Applicators need special (and often expensive) protective equipment, such as a self-contained breathing apparatus (SCBA), gas leak detectors, and more.
- Applicators of fumigants require more technical skill.
- Labor costs can be high.
- Fumigants must be retained in the gas form for a required period of time to be effective, often calling for extra supervision.

Physical and Chemical Characteristics of Fumigants

To understand how fumigants work, you need to understand certain physical and chemical characteristics of fumigant molecules. These characteristics determine how quickly a gas will spread to fill a space you are fumigating, how the gas might damage certain objects in the fumigation space, and more. Some of these characteristics include:

- *Molecular weight* and *specific gravity*,
- *Volatility* and *vapor pressure*,
- *Boiling point*,
- *Solubility*,
- *Flammability*, and
- *Chemical reactivity*.

> **Molecular Weight Units**
> Molecular weight is expressed as grams per mole (g.mol). This study guide does not cover how molecular weight is determined and the significance of the unit of measure. For simplicity, the molecular weight is presented as a number in this guide.

The characteristics above are different for each fumigant. Each characteristic makes a fumigant act a certain way under certain conditions. Understanding how these characteristics affect an application will help you select the best product for a particular job and, just as importantly, help you make a safe application.

Molecular Weight and Specific Gravity
Molecular Weight
All substances regardless of their state (solids, liquids, and gases) have a "molecular weight." Molecular weight is a measure of the weight of the atoms that form a particular substance. For example, table salt has a molecular weight of about 58. More complex molecules have greater molecular weight because they have more atoms (e.g., table sugar has a molecular weight of about 342).

Specific Gravity
All substances also have a "specific gravity." Specific gravity is the ratio of the density of a substance to the density of another substance used as standard. Water is usually the standard when comparing solids or liquids. For gases, including fumigants, air (specific gravity = 1) is used as the standard. Fumigants with a specific gravity greater than 1 are heavier than air and can sink. Fumigants with a specific gravity less than 1 are lighter than air and can rise.

The molecular weight and specific gravity of a fumigant can tell you how well it will distribute in a fumigation site. Most fumigant gases are heavier than air. For example, the specific gravity of sulfuryl fluoride is 3.52, indicating that it is 3.52 times heavier than air. When fumigants are heavier or lighter than air, the label will tell you if you need to use fans and/or other methods to evenly distribute the active ingredient during fumigation.

Volatility and Vapor Pressure
The terms "volatility" and "vapor pressure" are often used interchangeably. Although they are related, there is a subtle, but important, difference between the two.

"Volatility" describes how easily a substance will vaporize or turn into vapor. It is a relative term, like saying tomorrow will be cooler or warmer. Those terms don't tell you anything about the temperature itself (plus, one person's "cool" might be another person's "warm"). In the same way, we can say that one compound is more, or less, volatile compared to another, but that doesn't indicate how much more or less volatile it is. This is where "vapor pressure" comes in. Vapor pressure is an actual measurement of a physical property—how strongly that substance will push against the atmosphere.

If you place a chemical in a closed container, some of it will vaporize. The pressure in the space above the liquid increases from zero and eventually stabilizes at a constant value, its vapor pressure (Figure 2).

Since vapor pressure determines the concentration that can be maintained during fumigation, materials with a high

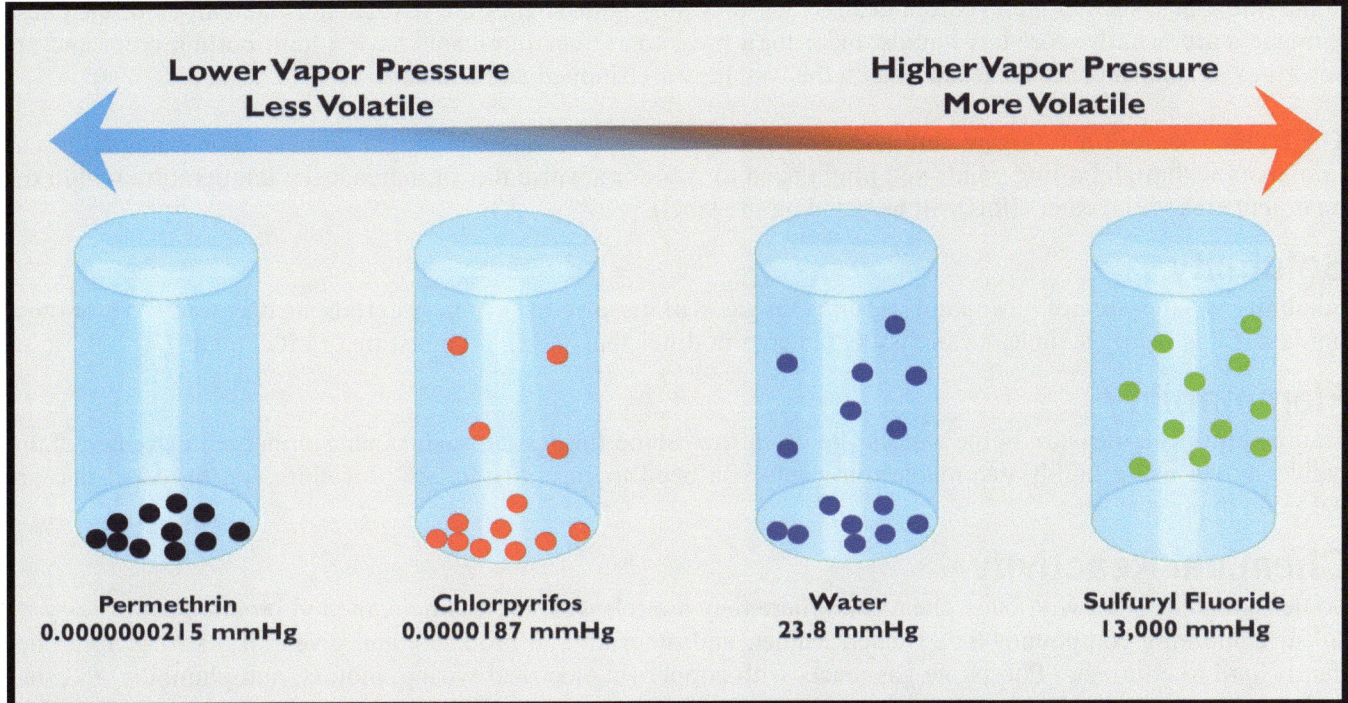

Figure 2: A comparison of the vapor pressure and volatility of some compounds. Above are some examples of compounds and their vapor pressures. Vapor pressure can be measured in different units, in this case they are given in millimeters of mercury (mmHg). The larger the number, the greater the pressure and the greater the tendency of the compound to turn into a vapor. Permethrin, an active ingredient used in some aerosol pesticide products, has a very low pressure and will barely volatilize. On the other end, the fumigant sulfuryl fluoride has a high vapor pressure and will easily volatilize and thus, vaporize. Vapor pressure is dependent on temperature. Generally, the higher the temperature, the greater the vapor pressure. (Illustration © UW-Madison Pesticide Applicator Training Program based on an illustration from the National Pesticide Information Center. npic.orst.edu).

vapor pressure are more concentrated and therefore have better fumigant qualities.

Volatility and Temperature

Volatility increases as temperature rises. Some fumigants, such as sulfuryl fluoride, exist as a gas at room temperature. Other fumigants (such as products that contain the active ingredients paradichlorobenzene or naphthalene and are used in moth balls and crystals) are liquids or solids at room temperature. Some of the "solid" fumigants, such as aluminum and magnesium phosphide, are not fumigants themselves but react with moisture to produce the fumigant gas phosphine (also called hydrogen phosphide).

Advantages and Disadvantages to Fumigant Volatilization

There are both positive and negative consequences to fumigants moving as individual molecules. On the plus side, fumigants can easily penetrate tiny spaces, diffuse into concrete, brick, and wood, or between the spaces of grains in a grain bin. This allows the fumigant to reach areas where pests might otherwise escape contact with the use of non-fumigant pesticides. The particles that make up aerosols are much larger (as noted above) and are unable to penetrate even a short distance into materials. For this reason, even though smokes, fogs, or mists are sometimes referred to as "fumigants," they are not true fumigants, and we will not deal with them further in this study manual.

On the negative side, the gaseous nature of fumigants can more easily cause harm beyond the treatment site. This is why it is so important to seal off treatment areas, monitor concentrations outside the area, post warning signs, and take other safety precautions that will be covered in later chapters.

Boiling Point

A boiling point is the temperature at which a chemical or fumigant becomes a gas. For example, the boiling point for water is 212°F. Boiling point is also related to the chemical's vapor pressure. In general, the higher the boiling point, the lower the vapor pressure, and the slower a fumigant will change to a gas.

Some fumigants, such as methyl bromide, have a low boiling point so they are gases at normal temperatures. These fumigants are usually stored as liquids under high pressure. Other fumigants have a high boiling point and are described as liquids or solids depending on the way they are shipped and handled.

The boiling point of a fumigant can determine the type of application equipment you use. For example, to use fumigants with high boiling points you might need to wait for a warm day or increase the temperature within the treatment area with heaters (this will be noted on the label).

Solubility

Solubility is a measure of how readily a fumigant gas will dissolve in certain materials such as water, oil, or other liquids. If it is highly soluble, it can dissolve in commodities that have high moisture or oil content.

Flammability

Flammability is a measure of the capacity to catch fire. Some fumigants, such as phosphine, are extremely flammable. When using highly flammable fumigants you need to carefully follow procedures on the label that are designed to prevent fires.

Chemical Reactivity

Some fumigants react with other chemicals where they are released. For example, methyl bromide combines with sulfur containing compounds (i.e., rubber, leather, and other animal products) and gives off a strong, foul odor that is hard to eliminate. Phosphine gas reacts with copper (in electrical wiring, motors, and plumbing) to cause serious corrosion. High temperatures around an open flame can cause some fumigants to form corrosive acids. Certain fumigants react with and damage photographic film and paper. The label will advise you of these potential reactivities.

SOME CHEMICAL CHARACTERISTICS OF COMMONLY USED NON-SOIL FUMIGANTS				
FUMIGANT	MOLECULAR WEIGHT	BOILING POINT (°F)	SPECIFIC GRAVITY	FLAMMABILITY (IN AIR)
Carbon Dioxide	44.01	-109.3	1.52	Nonflammable
Sulfur Dioxide	64.06	50.0	2.26	Nonflammable
Metam Sodium	129.18	230.0	1.16	Flammable (as MITC)
Methyl Bromide	94.95	38.4	3.28	Nonflammable
Phosphine	34.00	-125.9	1.17	Flammable
Sulfuryl Fluoride	102.07	-68.0	3.52	Nonflammable

Figure 3: Demonstrating diffusion with food coloring and water. The same principle involved in the diffusion of food coloring in water also applies to how a fumigant diffuses in air. Diffuse (verb): to cause a gas or liquid to spread through or into a surrounding substance by mixing with it. Diffusion (noun): a process by which there is a net flow of matter from a region of high concentration to a region of low concentration resulting from random motion of molecules. (Illustration © UW-Madison Pesticide Applicator Training Program).

Movement of Fumigants Through the Application Site

The movement of fumigant molecules released at the application site is determined by many of the chemical properties we just discussed and by *sorption*. Once released, fumigant molecules spread in the space they are contained in (e.g., building, burrow, fumigation chamber, tarped site, etc.) until they are evenly distributed in the air, this process is known as *diffusion* (Figure 3). The speed that the molecules move through air is directly related to their molecular weight. Heavier gases *diffuse* more slowly, and it may be necessary to help move these gases with fans or blowers. The rate that fumigants diffuse is also influenced by temperature: the warmer the temperature, the faster the diffusion.

Sorption

When a fumigant gas contacts materials, gas molecules undergo the process of sorption. There are two types of sorption: adsorption and absorption (Figure 4).

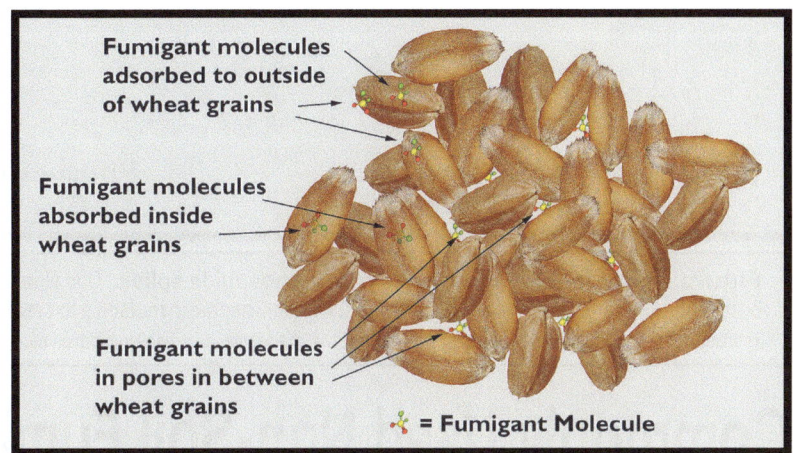

Figure 4: Adsorption and absorption. Both adsorption and absorption reduce the effectiveness of fumigants. These bits of fumigant are no longer able to move freely and kill the target pest(s). Sorption also slows aeration, the process of releasing the fumigant after treatment. (Illustration © UW-Madison Pesticide Applicator Training Program).

Adsorption occurs when fumigant molecules stick to a surface. That surface could be a treated material, such as grain, or the structure being fumigated itself.

Absorption occurs when the molecules penetrate a material (i.e., the treated product or the structure).

Desorption is the opposite process; it is the release of gas molecules from a material or surface to which the gas has attached itself.

Sorption Effects on Fumigation

Some fumigants are more likely to go through the process of sorption than others. Commodities and the structures that house them also vary in their sorptive capacity. For example, loads of grain with many small pieces have a lot of surface area and are more sorptive. Inert surfaces such as metal are less sorptive (but porous concrete may be more sorptive).

Knowing how sorptive certain surfaces and commodities are is critical. When treatment is complete, *aeration* periods must be long enough to allow the fumigant to desorb from the commodity. If aeration is too short, traces of fumigant could remain on the product causing toxic residues, off-flavors, and/or odors in the treated material. As a general rule, sorption is greater at cold temperatures.

> **IMPORTANT!** California's pesticide laws and regulations are stricter than many other states. Fumigants listed in this section or mentioned anywhere in this study guide are provided merely as examples. Always confirm that a fumigant is registered in California for use in the state, and on the specific commodity or site you intend to fumigate. Many of the non-soil fumigants require a restricted materials permit from the local County Agricultural Commissioner's (CAC's) office prior to use. If in doubt, consult with the CAC's office before using any fumigant for any purpose. Restricted Materials and restricted material permits are discussed in Chapter 2.

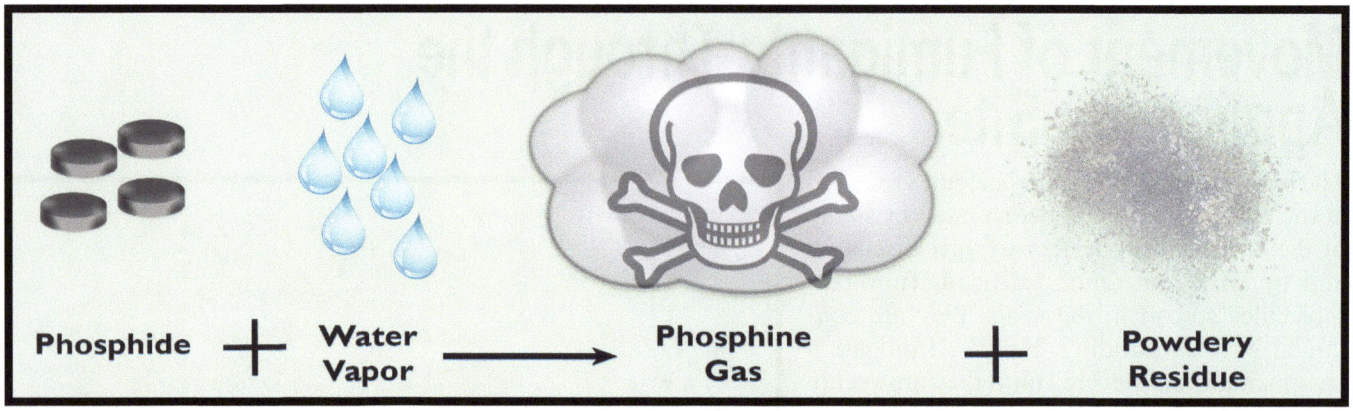

Figure 5: Phosphine gas production from phosphide solids. This illustration shows the reaction of phosphide solids (applies to both aluminum and magnesium phosphides) with atmospheric moisture to create phosphine gas. (Illustration © UW-Madison Pesticide Applicator Training Program based illustration by Ruth O'Neill, MSU in the Montana non-soil fumigation manual).

Commonly Used Non-Soil Fumigants

Some of the commonly used non-soil fumigants are:
- Phosphine
- Sulfuryl Fluoride
- Carbon Dioxide (CO_2)
- Methyl Bromide
- MITC and MITC-generating fumigants (Metam Sodium and Dazomet)
- Sulfur Dioxide (SO_2)

Phosphine

Phosphine is a toxic gas that can be generated from solid formulations of aluminum phosphide or magnesium phosphide. These formulations react with moisture in the air to produce the toxic gas phosphine (Figure 5). Magnesium phosphide releases phosphine much more rapidly than aluminum phosphide. There are also formulations of phosphine gas in pressurized cylinders either as pure phosphine or phosphine mixed with CO_2.

Phosphine is about 20% heavier than air but does not tend to *stratify*. Therefore, in open spaces you do not usually need fans to ensure even distribution of the gas. However, in bulk commodity fumigations, phosphine will not settle downward through grain without the use of fans. Once phosphine penetrates the grain, it diffuses rapidly because it is not strongly absorbed by grain. The combination of the low absorption of phosphine by the grain and the great penetration capacity of phosphine results in the tendency of the fumigant to leak if the structure holding the grain is not gas tight.

Choosing Solid vs. Gas Formulations

The decision to use solid formulations (aluminum and magnesium phosphide) instead of cylinders of phosphine gas will depend on several factors including the fumigation site, the skill of the applicator and, of course, the label.

Solid aluminum and magnesium phosphide comes in the form of tablets, pellets, ropes, strips, and more. There are benefits and drawbacks associated with the use of solid phosphide forms or pressurized cylinders of phosphine gas. Some of these include:
- Cylinders are basically bottled phosphine, so the target fumigant concentration is reached as soon as the gas spreads through the fumigation site. In contrast, packaged phosphide has about a 24 hour ramp-up period, where the solids need to react with humidity in the air to convert into phosphine gas.
- In a tightly sealed structure, cylinders of phosphine work quickly. However, poorly constructed structures lose gas quickly resulting in insufficient concentrations of the gas necessary to control the pest over the duration of

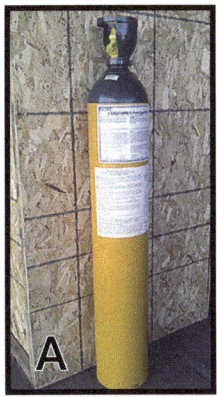 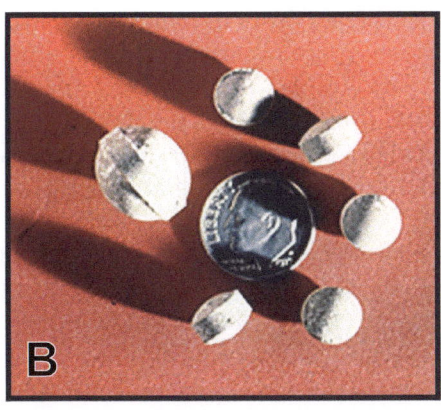 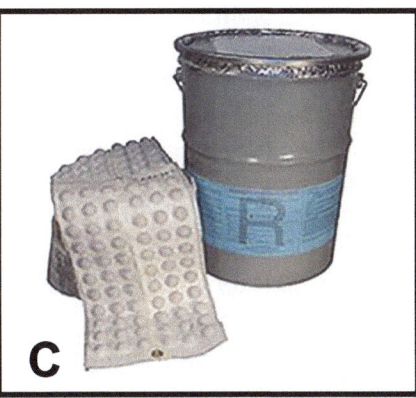 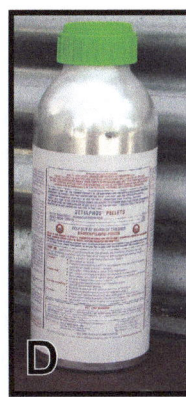

Figure 6: Phosphine packaged as phosphide solids (A, B and C), and as phosphine gas (D). (A) Since metal phosphide fumigants react readily with moisture, they must be packaged in gas tight metal containers. (B) Small containers usually contain loose pellets or tablets. (C) Another form the fumigants come in is as Prepac bags (also ropes) where the tablets are contained in gas permeable blister packs. Both Prepacs and ropes are then packed in metal containers where they must stay until use. Once removed from their sealed container they start producing phosphine gas after a short delay. Other formulations may start producing gas immediately. (D) Phosphine can also be applied from pressurized cylinders that contain phosphine, which is sometimes mixed with liquid carbon dioxide. (Photo credits: (A) © Betsy Danielson, Iowa State University Extension and Outreach; (B-D) © Cardinal Professional Products).

the fumigation.
- The primary advantage of solid formulations over cylinder applications is the cost.
- Smaller packaging is available for solid formulations, which is beneficial for small fumigations like boxcars, containers, fumigation chambers, and storage bins. Many manufacturers offer specifically sized packaging to make it easier to calculate the correct dosages for some fumigation sites.

Odor
Phosphine is a colorless gas that is odorless in its pure form, but technical grade phosphine sometimes has an odor described as "fishy" or "garlicky." Because that odor can vary, odor **does not** provide a reliable warning of hazardous concentrations and **should not be used** as a reliable indicator of phosphine's presence.

Uses and Applications
Both aluminum and magnesium phosphide fumigants are available in several different solid product forms, including pellets, tablets, Prepacs, bags, and plates (Figure 6). Phosphine can also be applied from pressurized cylinders directly as a gas either as pure phosphine or mixed with CO_2, which makes the mixture non-flammable.

The equations for the chemical reaction of aluminum phosphide (AlP) or magnesium phosphide (Mg_3P_2) with water (H_2O) to create phosphine gas (PH_3) are:

$$\text{Aluminum Phosphide: } AlP + 3\ H_2O \rightarrow Al(OH)_3 + PH_3$$
$$\text{Aluminum Phosphide: } Mg_3P_2 + 6\ H_2O \rightarrow 3\ Mg(OH)_2 + 2\ PH_3$$

When aluminum or magnesium phosphide are used, the concentration of phosphine increases slowly after first application—12 to 48 hours might elapse before the desired concentration is reached. The reaction of phosphides to moisture starts slowly, gradually accelerates, but tapers off again as the metal phosphide is spent and there is little to no metal phosphide left to react. The initial slow reaction is important to allow sufficient time for the applicator to open the package, place the fumigant, and vacate and seal the space. Magnesium phosphide releases phosphine much more rapidly than aluminum phosphide. Magnesium phosphide also releases the largest amount of phosphine which makes it better suited for fumigations conducted under cooler and drier conditions. Aluminum phosphide tablets are larger and release more phosphine than pellets, however, the pellet form reacts to create phosphine gas somewhat faster than tablets.

The total time needed for a successful fumigation can range from 3 to 10 days depending on temperature and other factors. Phosphine will be produced more quickly, and the fumigation completed sooner, under warm, humid conditions compared to cool, dry conditions. Generally, chemicals react faster as temperature increases. Since the production of phosphine requires moisture, higher humidity means more water is available to increase gas generation.

Phosphine gas product labeling may allow fumigations in cold storage rooms (as low as 28°F). In contrast, aluminum or magnesium phosphide fumigations will usually be ineffective at temperatures below 40°F. This is because gas release is slower, and insects breathe slower and do not inhale enough of the fumigant to obtain a fatal dose.

Phosphine and aluminum and magnesium phosphide fumigations are widely used for treating commodities in bulk storage and in transit (e.g., boxcars and ships), partly because they have no adverse effects on seed germination at label specified dosages. However, goods undergoing fumigation with aluminum phosphide in a truck trailer or container, or that have not been completely aerated, cannot be transported over public roads or highways. Doing so is illegal.

IMPORTANT! When aluminum phosphide or magnesium phosphide are used for outdoor burrowing rodent control, applications are prohibited within 100 feet of any building where humans and/or domestic animals do, or might, reside. Labels of products that are registered in California for burrowing rodent control will contain this, or a similar, statement: "For burrowing rodent applications: The use of this product is strictly prohibited within 100 feet of any building where humans and/or domestic animals do or may reside on single and multi-family residential properties, and nursing homes, schools (except athletic fields), day care facilities, and hospitals."

Both aluminum phosphide and magnesium phosphide are also used to control burrowing mammals. Phosphine, aluminum phosphide, and magnesium phosphide are restricted use pesticides and California Restricted Materials. A restricted materials permit from the local CAC's office is required prior to the purchase and use of these pesticides. Restricted materials and restricted materials permits are discussed in detail in Chapter 2.

Precautions

Fire and Explosion: A fire or an explosion can occur if phosphine is produced too rapidly. To reduce this risk, manufacturers produce controlled-release forms that help prevent explosion risk. Some solid formulations also give off CO_2, which helps minimize the risk of a fire or an explosion. Phosphine in pressurized cylinders mixed with CO_2 is not flammable.

Figure 7: Sulfuryl fluoride packaged in pressurized cylinders. Note the applicator is wearing eye protection while opening the cylinder. (Photo © Betsy Danielson, Iowa State University Extension and Outreach).

Explosions can occur if aluminum phosphide is used improperly. For example, phosphine is explosive at concentrations above 18,000 ppm. If the fumigation is conducted properly, concentrations should never approach this level. Phosphine is also explosive under vacuum conditions and should never be used in a chamber fumigation. Also note that aluminum phosphide can explode if it is in contact with water for extended periods.

The risk of fire or explosion described for aluminum phosphide also applies to magnesium phosphide. Magnesium phosphide is produced as pellets, tablets, or impregnated onto polyethylene plates and strips covered by a gas permeable paper. Magnesium phosphide is also packaged in small, gas permeable blister packs that can only be used for fumigating equipment.

Chemical Reactivity: Phosphine gas, especially in the presence of moisture or ammonia, reacts with silver, copper, and copper alloys. Copper containing equipment, such as electrical devices, can be severely damaged. Prevent electronic equipment from being exposed to phosphine by protecting or removing them from the application site.

Residues: Both aluminum phosphide and magnesium phosphide produce residue when they react with moisture in the air. Partially reacted dust is called "green dust" because it remains slightly greenish in color, while the spent residual dust is grayish white. Do not assume that a fumigation is complete just because you see residue, as gas may still be releasing even if residue is present. This residue should not ever come in direct contact with any processed food or bagged commodity. You can prevent the residue from contacting a commodity by placing the phosphides on a tray instead of adding it directly to the commodity. Special packages are available that retain the residue, preventing contamination. Residues must be completely deactivated and disposed of properly according to the label. At the end of the fumigation, you must collect all spent or partially spent aluminum and magnesium phosphide plates, strips, ropes, or Prepacs from the treated site.

Sulfuryl Fluoride

Sulfuryl fluoride is a colorless, odorless gas (Figure 7). It is nonflammable in all atmospheric concentrations but will become *corrosive* in the presence of an open flame, electric heaters, or other high heat sources. Fans or blowers are used to introduce, distribute, and aerate sulfuryl fluoride. Aeration is quite rapid, and the gas desorbs relatively quickly.

Uses and Applications

In California, sulfuryl fluoride is used to fumigate food and non-food commodities in vehicles, ships, rail cars, and fumigation chambers. Always check the product label to verify use site and commodity information. This fumigant is very effective at label specified concentrations against insect larvae and adults, but not eggs. The Structural Pest Control Board issues the license for the use of sulfuryl fluoride to control structural pests (e.g., termites, powderpost beetles, bedbugs, cockroaches, rodents, etc.). This study manual does not cover this use.

Sulfuryl fluoride is a restricted use pesticide and a California Restricted Material. Uses of sulfuryl fluoride covered in this study manual require a restricted materials permit from the local CAC's office prior to the purchase and use of this pesticide.

Precautions

Sulfuryl fluoride can decompose into sulfur dioxide (SO_2), hydrofluoric acid, and other decomposition products. Hydrofluoric acid is highly reactive and can corrode or damage many materials.

Chemical Reactivity: Sulfuryl fluoride is noncorrosive and unreactive to most materials. However, in the presence of an open flame and glowing heat elements, it forms a very corrosive gas. Turn off all pilot lights and allow heating devices to cool to prevent formation of a

Figure 8: The only structural fumigation use for chloropicrin is as a warning agent with sulfuryl fluoride fumigations. You can do this by pouring a small amount of chloropicrin onto some cotton in a shallow pan placed in front of a fan (Top). The bottom picture shows a close up of the chloropicrin pan. Allow 5 to 10 minutes for the chloropicrin to circulate before introducing the fumigant. (Photo credits: (top) © Bonnie Rabe and (bottom) © Betsy Danielson, Iowa State University Extension and Outreach).

corrosive gas. To address this hazard, labels may also direct you to shut off automatic switch controls for appliances and lighting systems. Additionally, if released too quickly from the cylinder, sulfuryl fluoride can condense as a liquid that can combine with water particles in the air to form small amounts of hydrofluoric acid, an acid that can cause etching of glass, metals, and other surfaces.

Chloropicrin

The main use of chloropicrin is as a *warning agent* with sulfuryl fluoride in structural fumigations. In California, the Structural Pest Control Board issues the license for fumigations to control structural pests (e.g., termites, powderpost beetles, bedbugs, cockroaches, rodents, etc.). This study manual does not cover this use.

The other notable non-soil fumigation use of chloropicrin is to fumigate in-service utility poles, pilings, and similar timber members to prevent internal decay. However, at the time of this writing, these uses are not registered for use in California. Therefore, this manual will not further discuss chloropicrin's characteristics.

Carbon Dioxide (CO_2)

Carbon dioxide is a non-flammable, colorless, and odorless gas that naturally occurs in the atmosphere. However, it can be used as a fumigant to kill pests when introduced at a sufficiently toxic concentration. Treatment with CO_2 involves displacing the air inside a gas tight structure with a concentration level of CO_2 high enough to be toxic to pests.

Uses and Applications

Carbon dioxide can be used to kill insect pests in stored commodities and for burrowing pest control. It does not leave residues on products or structures. CO_2 is applied from pressurized tanks. Fumigations with CO_2 need to take place in fumigation chambers or in tightly sealed structures.

Carbon dioxide can kill all life stages of insects, but generally needs a long treatment period. Also, the CO_2 needs to be held at a concentration of 35% to 60% at room temperature for up to 21 days. The concentration of CO_2 and exposure time will differ depending on the pest, commodity type, and temperature.

Precautions

Although CO_2 is naturally found in the air we breathe, it is present at a very low concentration (0.04% or about 400 ppm). Because of the concentrations required to effectively control pests, CO_2 used as a fumigant is lethal to humans.

Methyl Bromide

Methyl bromide is a colorless, odorless gas. For fumigant use, it comes in pressurized cylinders as a liquid. When released, it vaporizes at temperatures above 38.5°F to form a gas that is 3.28 times heavier than air. Recirculation is used to ensure even distribution of the fumigant. Being heavier than air, methyl bromide tends to settle out in low places. It also tends to stratify, so fans are needed to assure thorough mixing of the gas with the air. It penetrates most commodities very well and is effective against all stages of insect life.

Uses and Applications

Methyl bromide was once widely used as an agricultural soil, commodity, and structural fumigant. However, most uses have been phased out under the federal Clean Air Act because methyl bromide depletes the Earth's ozone layer. The fumigant is still allowed for quarantine fumigations under the oversight of the U.S. Department of Agriculture–Animal and Plant Health Inspection Service (USDA – APHIS) and for a limited number of critical use exemptions.

Methyl bromide is applied in structures containing raw products such as warehouses, grain elevators, and food processing plants. It is also applied to boxcars, ships, trailers, and can be used in tarp fumigation.

> **Methyl Bromide Phaseout**
>
> Under the Clean Air Act, most uses of methyl bromide have been phased out because it contributes to depletion of the Earth's ozone layer.
>
> There are some exceptions to the phase out called Critical Use Exemptions (designed primarily for agricultural users without any technically or economically feasible alternatives), and some Quarantine and Pre-shipment exemptions. There are additional documentation requirements for exempted uses. If you can use the product under one of the exemptions, be sure you are properly trained in the use of the methyl bromide and read the product label (as you should for all pesticides) before using the fumigant. For more information on the methyl bromide phaseout and exemptions, refer to EPA's website.

Methyl bromide is a restricted use pesticide and a California Restricted Material. A restricted materials permit from the local CAC's office is required prior to the purchase and use of this pesticide.

Precautions

Fire and Explosion Risk: Although there is no fire hazard at application rates indicated on the product label, you must assure that all pilot lights and other open flames are turned off prior to use. In the presence of an open flame and glowing heat elements, methyl bromide forms a highly corrosive gas.

Chemical Reactivity: Numerous items and materials should not be exposed to methyl bromide since the gas may directly damage these materials. For example, never use rubber products (such as rubber hoses) with methyl bromide. Methyl bromide combines with sulfur-containing compounds (such as rubber, leather, and other animal products) and gives off a strong, foul odor that is hard to eliminate. Some materials that should not be fumigated with methyl bromide include iodized salt, materials that may contain reactive sulfur compounds (such as some soap powders, and some baking sodas), sponged rubber, foam rubber (as in rug padding, pillows, cushions, and mattresses), leather goods, and woolens. Read the label for a complete list of materials that may either react with methyl bromide or should not be exposed to methyl bromide.

Methyl bromide might damage seeds and bulbs that are to be used for planting, living plants, or nursery stock. Under most conditions, fumigation with methyl bromide will adversely affect seed germination. The use of methyl bromide is not recommended if moisture is high or if temperatures are below 40°F.

MITC and MITC-generating Fumigants (Metam Sodium and Dazomet)

Methyl isothiocyanate (MITC) is a broad-spectrum fumigant which is toxic to most life forms. Although generally more toxic than methyl bromide, MITC does not volatilize or disperse as well as other fumigant active ingredients. MITC itself is a colorless solid with a melting point of 88°F.

MITC is typically applied as an "MITC generator." An MITC generator is a chemical that, upon exposure to the environment, degrades to form MITC gas shortly after application. The MITC gas acts as the fumigant, killing the target pests. Examples of MITC generators include metam sodium (Sodium N-methyldithiocarbamate) and dazomet (Tetrahydro-3,5-dimethyl-2H-1,3,5-thiadiazine-2-thione), both are broad-spectrum fumigants. Metam sodium is a white crystalline powder in its pure form but is usually found as a clear yellow-green liquid with a sulfur odor that can vary in intensity. It is a contact, non-systemic pesticide. When metam sodium breaks down, the breakdown products, in addition to MITC, include carbon disulfide (CS_2), hydrogen sulfide (H_2S), elemental sulfur, and 1,3-dimethylthiourea. When dazomet breaks down, the breakdown products are formaldehyde, monomethylamine, H_2S, and in certain situations, CS_2.

Uses and Applications

Metam sodium and dazomet are used frequently as soil fumigants. Metam sodium's non-soil fumigant uses include wood preservation of utility poles, in leather processing, in raw cane and sugar processing, antimicrobial treatments in sludge and industrial water purification systems, and as an herbicide for sewer root control (commonly in combination with dichlobenil).

Dazomet's non-soil fumigant uses include use as bactericide and fungicide in industrial processes and water systems, for material preservation, and for wood preservation. The industrial uses include oil fields, paper mills, and recirculating cooling water systems. The material preservation uses include clay slurries, adhesives, coatings and high viscosity suspensions, construction materials, epoxy flooring compounds, paints, inks and dyes, as well as paper auxiliaries and additives. In wood preservation, dazomet is used in remedial treatment of utility poles, pilings, and timbers.

Precautions

Metam sodium is stable under normal conditions and very stable at a pH of 8.8 or greater, but unstable below pH 7. Prolonged exposure to air results in gradual decomposition to the poisonous gas MITC. MITC is much more toxic than metam sodium and can reach unsafe levels in poorly ventilated or confined spaces. When metam sodium is mixed with water it quickly converts to MITC gas. MITC gas can be used to kill roots by penetrating the root system. The use of MITC for sewer line root control is discussed in detail in Chapter 7.7.

Flammability and Reactivity: Metam sodium itself is not flammable but its breakdown product, MITC, is. Metam sodium solutions are corrosive to copper, zinc, and aluminum or any of their alloys such as brass, bronze, or galvanized metals.

Sulfur Dioxide SO_2

Sulfur dioxide is a non-flammable, whitish colored gas with a pungent odor. Its boiling point is 50°F. It is corrosive in the presence of water. Sulfur dioxide can be sensed by taste at low level concentrations.

Uses and Applications

Sulfur dioxide products are used as a postharvest treatment for grapes held in cold storage in enclosed spaces (e.g., warehouses and transportation vehicles) to control gray mold disease caused by the fungus *Botrytis cinerea*. SO_2 is also used as a fumigant in the wine industry. The wine industry uses SO_2 to sanitize barrels and corks to prevent microorganisms (bacteria and fungi) that may be present in barrels and corks from contaminating wine and causing it to spoil or off flavor.

Precautions

Sulfur dioxide is acutely toxic to the respiratory system and can potentially harm workers and/or bystanders. The peak effect occurs within 5 to 10 minutes of exposure at relatively low concentrations.

Avoid breathing SO_2 gas as it can be fatal if inhaled in high concentrations. Anyone with a history of respiratory problems should avoid any exposure. SO_2 is an eye, nose, and throat irritant even at low levels. If tearing or upper respiratory tract irritation symptoms occur, leave the fumigation area immediately.

Product labels may suggest a medical assessment of applicators and other persons who will be regularly exposed to SO_2. These assessments are conducted prior to employment and at 1 to 2 year intervals.

Pest Factors That Influence Fumigation

Pest Identification

Accurate pest identification is the key factor in determining which control method or methods will be used to address the pest problem. Pests vary in their response to control methods and accurately identifying the target pest is extremely important as it can make the difference between effective pest control and failed control. Aside from correctly identifying the target pest, knowing its biology, including life cycles, is also key to effective pest management. This study manual will cover in more detail the biology of termites (Chapter 8.1), wood-boring beetles (Chapter 8.2), stored commodity pests (Chapter 8.3) and other miscellaneous pests (Chapter 8.4)."

Pests Not Controlled with Fumigation

Fumigants tend to affect all forms of life. Almost any pest in an enclosed area can be killed if exposed to the appropriate concentration of a fumigant. It is very important to understand the habits of the pest before choosing fumigation as a control method. A fumigation will only kill the pests inside the treated area or space. A pest problem outside the treated area serves as a reservoir of the pest population that can quickly reinfest the treated area. Rodents or insects that have easy access to the fumigated area can also reinfest an area after a fumigation. This can be prevented by making sure all available entry points are properly sealed so the pests cannot enter (or re-enter) in the first place.

Different life stages of the pest might also respond differently to fumigation. For example, many insects are less susceptible to fumigants or other insecticides during their egg and pupal stages. Insects that are dormant during certain periods are also less susceptible because their respiration is slower and they take in less of the fumigant.

Pest Density

High pest densities and/or targeting multiple pest species may require using higher fumigant application rates but you should never exceed rates indicated on the label.

Application Rate and Timing

The biology of the pest can influence the effectiveness and timing of a fumigant application. Some biological factors include the:
- **Pest's stage of growth:** Adult and immature insects are generally the easiest life stages to kill. Eggs and pupae

are more difficult to kill and fumigants will vary in how well they kill these life stages.
- **Pest's activity level:** Active immature and adult insects are easier to kill because they respire more and take in more fumigant. Their higher metabolism also results in processing the fumigant faster.
- **Size of the infestation:** Smaller infestations are easier to control. Large masses of pests may generate dust, damaged grain, webbing, or cast skins that interfere with fumigant penetration.

Pesticide Resistance

Just like other pesticides, pests can develop resistance to a fumigant's toxic effects. Resistance to fumigants has emerged in many parts of the world. Some species of stored product pests, particularly the lesser grain borer, have developed resistance to several fumigants. In the U.S., phosphine resistance has been reported in red and confused flour beetles, the lesser grain borer, almond moth, cigarette beetle, Indian meal moth, rusty grain beetle, and, most recently, the sawtoothed grain beetle. A factor that contributes to the development of pest resistance is applications of the fumigant at very low and ineffective concentrations. Once a pest population develops resistance to the specified label rates of a fumigant, it becomes hard or impossible to control with that fumigant.

The same methods of resistance management used for other pesticides also apply to fumigants. Incorporating the following practices helps reduce pesticide resistance:
- Practice Integrated Pest Management (IPM): combine all available control measures into a practical pest management program.
- Use fumigants only when necessary: a pest population will develop resistance only when you use that fumigant against it. If you use the fumigant even when you don't need to, you may unnecessarily increase the proportion of resistant individuals.
- When using fumigants, rotate chemistries whenever possible. For example, sulfuryl fluoride might be used at intervals in a management program that relies mainly on phosphine. However, for some fumigations (such as a quarantine fumigation) there may be only one registered or acceptable fumigant, so there is no option for switching.
- Always follow the dosage rates and exposure times specified on the label.

> **How Pesticide Resistance Happens**
>
> Some individuals in a pest population (say, a group of insects in a grain bin) have genetic traits that allow them to survive a pesticide application. When sprayed with a pesticide those individuals will survive and the survivor's offspring will inherit the resistant traits. This will mean a larger population of resistant individuals in the succeeding generations. And if continued selection is applied by using the same pesticide (or different pesticides with the same mode of action), eventually a population consisting mostly of resistant types will come about.
>
> **Phosphine Resistance Testing**
>
> Some fumigation companies provide testing for phosphine resistance or kits you can purchase to do your own resistance testing.

Fumigant Resistant Issues

There are some issues with fumigants that can contribute to the occurrence of pest resistance:
- There are a limited number of fumigants to use for non-soil fumigations. Rotating the use of different fumigants may not always be an option.
- Quarantine protocols might force the use of fumigants over other pest control options, or only allow a certain fumigant to be used.
- Even low numbers of pests can cause severe consequences when transported from one part of the world to another, making total eradication of pests in transit a high priority. Pests that survive after treatment will produce offspring that are also hard to kill and can withstand above average doses of the fumigant.
- In a grain bin, on a cargo ship, or any other place where a resident population of insects is treated repeatedly with the same fumigant, resistance might develop.

The Safe and Effective Use of Pesticides, 3rd Edition, Chapter 10 (Whithaus, 2016), details further the topic of pesticide resistance.

When a Fumigation Goes Wrong
Real World Examples

What Happened?
An unlicensed seed warehouse applicator placed Fumitoxin® (aluminum phosphide) pellets inside cloth sacks to fumigate seed. Afterwards, the applicator put the sacks, with partially spent pellets, into the garbage. After the garbage was picked up, the garbage truck driver discovered his load was on fire, so he returned to the yard and dumped the load on the cement and contacted the fire department. The fire department arrived, and applied water, which made things worse. They discovered the source of the reactivity, the cloth sacks that would pop and smoke when water hit them. Squatting alongside the sacks, without wearing self-contained breathing apparatus (SCBA), they pondered what might be inside. It wasn't until a sheriff's deputy retraced the driver's route and determined the material as Fumitoxin®.

What Should Have Happened?
The seed company should have hired a licensed applicator to develop a fumigation management plan and conduct the application. A licensed applicator would have properly deactivated the partially spent product as per directions in the applicator's manual. If aware of the mishandling, the seed company should also have notified the local fire department, allowing the fire department to be notified of the potential hazard of the situation in advance.

How It Could Have Been Avoided?
Following the label and applicator's manual and having a trained and certified applicator.

What Were the Consequences?
Eleven individuals sought medical treatment. The seed company and the applicator were cited by the Washington State Department of Agriculture.

Chapter 1 Review Questions

Correct answers are given on page 175.

1. A substance that vaporizes readily is _____.
    a. volatile
    b. corrosive
    c. soluble

2. Which fumigant reacts with water to form a gas?
    a. magnesium phosphide
    b. carbon dioxide
    c. sulfuryl fluoride

3. A key component of the fumigation process is _____.
    a. pest sterilization
    b. pest preservation
    c. pest identification

4. The fumigant methyl isothiocyanate (MITC) is a breakdown product of _____.
    a. methyl bromide
    b. metam sodium
    c. sulfur dioxide

5. What conditions would increase the reaction rate of phosphine?
    a. warm and humid
    b. cold and dry
    c. warm and dry

CHAPTER 2

Fumigant Labels and Fumigant Handling

LEARNING OBJECTIVES

- ☑ Describe the importance of a pesticide label and its legal implications for an applicator.
- ☑ Define the terms "label" and "labeling."
- ☑ Describe fumigant-specific information found on the label.
- ☑ Outline the procedures for transporting fumigants.
- ☑ Describe the proper storage of fumigants.
- ☑ Describe how to respond to fumigant spills, leaks, exposure, or accidents.
- ☑ Explain how to safely dispose of leftover fumigant and fumigant containers.

Terms to Know

The following are important terms to know from this chapter. They will be *italicized* in their first use. They are explained in the text and defined in the glossary at the end of this manual.

Applicator's Manual	Restricted-Use Pesticide	County Agricultural Commissioner	Agricultural Commodity
Label			Property Operator
Labeling	Restricted Material	Restricted Material Permit	Agricultural Use

California Pesticide Use Requirements

Pesticide applicators must understand the laws and regulations and pesticide labeling before applying any type of pesticide, including fumigants. Federal and state laws and regulations as well as local use requirements must be followed. Always follow the strictest requirements when preparing to conduct a non-soil fumigation.

Business Licensing Requirements

If you or your employer advertises, solicits, or operates a business to conduct pest control for hire in California, the business must be licensed by DPR as a Pest Control Business. In California, the Structural Pest Control Board, not DPR, licenses the use of non-soil fumigants, such as sulfuryl fluoride, to control structural pests (e.g., termites, powderpost beetles, bedbugs, cockroaches, rodents, etc.) in structures.

Each Pest Control Business location (main and branches) must have an individual with a valid Qualified Applicator License who is responsible for the pest control operations of the location. The responsible qualified applicator must possess the pest control categories covering the pest control work to be conducted by the business. The qualified applicator may not supervise more than one Pest Control Business location.

Restricted Materials and Permitting

Title 3 of the California Code of Regulations (3 CCR) section 6400 designates the following pesticides as restricted materials:
- Any pesticide labeled as a Restricted Use Pesticide;
- Any pesticide used under a Section 18 Emergency Exemption; and
- Certain other pesticides listed in the section.

Some non-soil fumigants (i.e., phosphine, aluminum and magnesium phosphide, sulfuryl fluoride, and methyl bromide) are listed in 3 CCR section 6400 and require a restricted material permit to possess or use the fumigant.

These permits are obtained from the local *County Agricultural Commissioner*'s (CAC)'s office.

Restricted material permits are categorized as either agricultural or non-agricultural use. It is important to understand this classification to know who can obtain a permit and what information the applicant must provide the CAC's office. California has a broad legal definition of "agricultural use." In addition to pest control on farms, forests, and nurseries, *"agricultural use"* includes pest control in or on areas such as golf courses, parks, and rights-of-ways (e.g., highway medians, canals, railroad shoulders). "Non-agricultural use" includes pest control conducted for home, industrial, institutional, or vector control uses. Structural use is considered "non-agricultural use" in California, however this use is exempt from a permit and is not covered in this study manual. Several of these terms are defined in 3 CCR section 6000.

A permit for the "agricultural use" of a restricted material is issued to the *property operator* to allow them to purchase, use, and store these products. Agricultural use permits are site and time specific and require the property operator to identify all known "sensitive sites." "Sensitive sites" are areas that could be adversely impacted by the use of the pesticide(s) specified on the permit application. This includes areas such as hospitals, schools, playgrounds, residential areas, labor camps, parks, lakes, waterways, wildlife management areas, livestock, or crops. Prior to applying for a permit, the grower must consider, and if feasible, adopt any reasonable, effective, and practical mitigation measure, or use any feasible alternative which would substantially lessen any significant adverse impact on the environment. The agricultural use permit will include the name of the business conducting the application and the name of the certified applicator responsible for the application or supervising the application of the product.

There are several important differences in the information required for a "non-agricultural use" permit. For example, a permit for the "non-agricultural use" of a restricted material is issued to the property operator, the pest control business, or both. Also, the permit application must identify the criteria for determining the need for the pesticide application. Another key difference is the qualifying individual for a non-agricultural permit needs to have a Qualified Applicator License or Certificate.

The CACs may establish local use requirements called *permit conditions*. These are stricter requirements than found on the pesticide's labeling or California law and regulation. Permit conditions are issued by the CAC and must be followed. If violated, an individual may be cited, fined, or referred to the district attorney for prosecution of a crime. If using a non-soil fumigant in more than one county, a restricted materials permit must be obtained from each county in which the application will be conducted within. Restricted material permits are typically issued for one year. In addition, the permittee must submit a Notice of Intent (NOI) to the local CAC. For example, the NOI for an agricultural use permit must be submitted at least 24 hours prior to commencing the use of a pesticide requiring a permit. In some instances, this notification may be longer than 24 hours. Check with the local CAC for specific details on NOI requirements for both agricultural and non-agricultural permits.

Pesticide Recordkeeping and Use Reporting

For those pesticides where a restricted material permit is not required, the property operator may still be required to obtain an operator identification number from the local CAC prior to the use of the pesticide (pest control businesses are exempt from this requirement).

For many pesticides, the applicator (whether property operator or pest control business) must maintain records of the pesticides they apply and submit a pesticide use report for the fumigation to the local CAC. The use reports for some non-soil fumigations (such as burrowing rodent fumigations in a field, vineyard, or orchard) which are applied to or for the production of an *agricultural commodity* are submitted on the production agriculture pesticide use report form for each application conducted. However, most non-soil fumigations are not for the production of an agricultural commodity. These applications are reported to the CAC as a monthly summary of the applications completed. For more information on operator identification numbers, recordkeeping, and use reporting, review the Laws and Regulations Study Guide, 3rd Edition, Chapter 4 (California, 2020), or contact your local CAC's office.

Fumigation of Enclosed Areas

California has specific requirements when fumigating enclosed areas such as vaults, chambers, greenhouses, vans, shops, vehicles, and tarp-covered commodities. In addition to product label requirements, the following must be

followed for enclosed area fumigations:
- At least two trained employees must be present at all times when:
    - A fumigant is introduced into an enclosed area [Exception: only one trained employee is required to be present if the product used is a solid fumigant and is introduced into the enclosed area from the outside],
    - The area is entered to undergo aeration, and
    - The area is entered to determine the concentration of the fumigant and PPE is required by product labeling or regulation.
- The second trained employee must have immediate access to the handler PPE required by the label.
- The appropriate warning sign(s) shall be posted in a plainly visible location on or within the immediate vicinity of all entrances of the area under fumigation, prior to the beginning of the fumigation.
- Employees shall not be allowed to enter the fumigated enclosed area, except to determine the fumigant concentration or facilitate aeration (unless the concentration is known and below the required levels for entry as stated by label or regulation).
- The fumigant shall not be released into occupied work areas.
- Upon conclusion of the enclosed area fumigation, the treated area, or products, shall be managed to ensure employees entering the area or working with treated products are not exposed to fumigant concentration levels in excess of those stated by label or regulation.

Labels and Labeling

The pesticide *label* is the main method of communication between pesticide manufacturers and pesticide users. It is one of the most important tools for their safe and effective use. Manufacturers are required by law to put certain information on the label. It is extremely important that you read, understand, and follow all the information on the pesticide label. Without it, you cannot use the product safely and effectively no matter how knowledgeable or experienced you may be.

 REMEMBER! The label is the law. The label (and any supplemental labeling referred to on the label) is a legally binding document. That means that you must follow label directions explicitly!

Before we discuss the information found on pesticide labels, we must differentiate the terms *label* and *labeling*. The label is the information printed on or attached to the pesticide container. Labeling includes the label itself plus all other information about the product referenced on the label and/or given when you buy the product.

Labeling

Labeling might include information that accompanies the product. Labeling can include a comprehensive applicator manual, brochures, leaflets, and more. Not all the information necessary to use the pesticide will be found only on the pesticide container itself but may instead be found in the accompanying labeling. This is especially true for fumigant products, which almost always have separate, comprehensive *applicator manuals* that you must follow to use the product safely and legally (Figure 1). To reiterate: **ALL information, whether on the label attached to the product or that the label refers you to, is legally binding!** For the remainder of this chapter and this study manual, "the label" refers to any information on the attached label or any other information the label refers to (i.e., applicator's manual).

Information on Labels

Just like the labels of other pesticide products you use, fumigant labels will have all the usual identification, safety, and use information you need to use the product correctly and safely. Below we will discuss the type of information that is commonly found on fumigant labels.

> **RESTRICTED USE PESTICIDE**
> DUE TO HIGH ACUTE INHALATION TOXICITY OF PHOSPHINE GAS
> For retail sale to Dealers and Certified Applicators only.
> For use by Certified Applicators or persons under their direct supervision, and only for those uses covered by the Certified Applicator's certification. Refer to the directions in the application manual for requirements of the physical presence of a Certified Applicator.
>
> ==THE COMPLETE LABEL FOR THIS PRODUCT CONSISTS OF THE CONTAINER LABEL AND APPLICATOR'S MANUAL WHICH MUST ACCOMPANY THE PRODUCT. READ AND UNDERSTAND THE ENTIRE CONTAINER LABEL AND APPLICATOR'S MANUAL. REFER TO THE APPLICATION MANUAL FOR DIRECTIONS FOR USE, PRECAUTIONS AND RESTRICTIONS.==
>
> A FUMIGANT MANAGEMENT PLAN MUST BE WRITTEN FOR ALL FUMIGATIONS PRIOR TO ACTUAL TREATMENT.
>
> CONSULT WITH YOUR STATE LEAD REGULATORY AGENCY TO DETERMINE REGULATORY STATUS, REQUIREMENTS, AND RESTRICTIONS FOR FUMIGATION USE IN THAT STATE.
>
> **VAPORPH$_3$OS® Phosphine Fumigant**
>
> Pure phosphine gas for on-site blending with registered or food grade carbon dioxide or forced air to produce a non-flammable fumigant gas for use in controlling pests in enclosed empty spaces and enclosed space (including temperature-controlled spaces such as cold storage chambers, transport containers and other suitable fumigation spaces) containing listed raw agricultural commodities, processed foods, stored tobacco, animal feeds, and nonfood products. Not for use on barges. Not for burrow treatment. For details refer to the VAPORPH$_3$OS® Phosphine Fumigant Application Manual.

Figure 1: The fumigant label and labeling. An example of a fumigant label that directs users to the product's applicator manual—a separate document the user must read and understand before using the product. The product label and the applicator manual are both legally binding documents.

Fumigant Label Information

- **Restricted Use Information:** Most fumigants are classified as *restricted-use pesticides* (RUPs). The restricted-use statement will always be found at the top of the first page of the label. Only certified applicators may purchase and apply RUPs. In California, all RUPs are also considered state Restricted Materials (RMs).

- **Site Information:** Make sure the site you wish to fumigate is listed on the product label. For example, some products may only be used in non-residential structures so using a fumigant in an apartment building, for example, would not be allowed.

- **First Aid Information:** All pesticides have first aid information on their labels, and you always need to be familiar with that information before using the product. Fumigants, by their nature, often have very specific first aid information you need to know.

- **Personal Protective Equipment (PPE):** Many fumigants have very specific PPE requirements, especially when it comes to respirator use. Some of the respirator requirements will depend on the activity you are doing, such as releasing the gas, checking air concentrations, etc. Other PPE directions are also product specific. For example, you should not wear rubber gloves or boots when using methyl bromide because the gas interacts with rubber. Respirators and PPE are covered in detail in Chapter 5, "Personal Protective Equipment."

- **Physical and Chemical Hazards:** Many fumigants have specific temperature requirements for storage and/or use. Some can react with and damage certain materials. Some are highly flammable, and you need to make sure all sources of flame or sparks (e.g., pilot lights, heating elements) are turned off before fumigation.

- **Environmental Hazards:** Some fumigant labels might state that the product is toxic to wildlife and that non-target organisms exposed to it will be killed.

- **Container Handling:** Fumigants come in many different containers, from pressurized cylinders containing gas to pellets, or tablets packaged in gas tight containers. It is very important to note the temperature requirements for storing the containers. Also, you need to be extremely cautious when transporting and opening pressurized containers. Labels will also specify how to dispose of containers.

- **Monitoring:** Depending on the fumigant, you might have to monitor the air concentrations for the chemical during and after the application. You also might need to monitor outside the application area to check for leaks. The label may specify the monitoring equipment to use (Figure 2).

- **Aeration:** Aeration procedures when the fumigation is completed will vary depending on the type of fumigant used and the application site. The label will specify proper aeration protocols.

> **VII. Gas Detection Equipment**
>
> At all times there must be more than one each of Phosphine and CO2 (if used) detection equipment capable of measuring at least < 0.3 ppm for Phosphine and < 5,000 ppm for CO2 present at the stationary and in-transit fumigation and aeration sites. They must be periodically checked for accuracy. There are a number of devices on the market for the measurement of Phosphine gas as well as carbon dioxide levels for industrial hygiene purposes. Glass detection tubes used in conjunction with the appropriate hand-operated air sampling pumps are a widely used method. These devices are portable, simple to use, do not require extensive training and are relatively rapid, inexpensive and accurate. Electronic devices are also available for both low level and high Phosphine and carbon dioxide gas readings. The newer low-level electronic units as well as the low-level detector tubes can detect 0.01 ppm of Phosphine and are suitable for industrial hygiene monitoring. Such devices must be used in full

Figure 2: Monitoring requirements on a fumigant label. An example on a fumigant label that specifies the correct monitoring equipment to use. It is extremely important to use the correct monitoring equipment since precise measuring is essential for both effective treatment, and to protect the health and safety of applicators and the public.

- **Pest Factors:** Labels give dosages and exposure times needed for different species of insects and the life stages of those species. Be sure to check if there are any minimum temperatures or exposure times associated with the life stage of the pest.
- **Number of Applicators:** Many labels will specify that more than one trained and certified applicator must be present during certain stages of the application. Some labels may require that anyone working with the product must be a certified applicator with the Non-Soil Fumigation category.

Transporting Fumigants

All fumigants are classified as hazardous materials by the U.S. Department of Transportation (DOT) and are subject to extensive transport regulations. Chemicals are listed as hazardous materials because they can pose an unreasonable risk to health, safety, and property if spilled during transport. When transporting such materials, you may need to:

- Receive hazardous material training,
- Carry emergency response information in your vehicle,
- Carry shipping papers in your vehicle,
- Placard your vehicle, or
- Have a Commercial Driver's License (CDL).

You also should have vehicle safety kits and inspection and maintenance logs. Refer to both federal and California DOT regulations to determine the placarding requirements for transporting each fumigant. Contact the fumigant manufacturer or distributor for more information on placarding for transportation (Figure 3).

Figure 3: Placards for transporting fumigants. An example of a placard that may need to go on the outside of your vehicle when transporting certain pesticides.

Fumigant labels give specific instructions for transport of their products. Always refer to the label to make sure you are correctly transporting the product.

Other Precautions

When transporting fumigants, use common sense and take the following precautions:

- Always transport fumigants in a separate air space from vehicle occupants.
- You must secure fumigant gas cylinders or other containers so they do not move during transport (Figure 4).
- Be sure you have the required driver's license and driving class with any appropriate endorsements for the specific fumigant you plan to transport.
- You must transport pressurized cylinders with the valve cover and safety bonnet attached.
- Do not remove protective valve covers until just before use.
- Always follow federal and California DOT regulations when transporting fumigants and/or their containers.

Storage of Fumigants

Many businesses store pesticides in existing buildings that have other uses. However, a separate building that is well-ventilated and/or has a mechanical exhaust system is best especially for fumigants. Make sure you can properly heat and cool the storage facility. Many fumigants have specific temperature requirements, which will be listed on their labels. Also, check the label to make sure there are no potential problems storing fumigants where there might be pilot lights or other heating elements that could be of concern. You must protect and secure (i.e., lock) the storage area to keep out unauthorized people (especially children) and animals. Some labels might require fumigants to be locked away when not in use. In addition to any label requirements, be sure to check with appropriate officials to ensure you are meeting any state, county, or local requirements for fumigant storage.

Pesticide Storage Posting

Make sure you properly placard all fumigant storage areas. The label and/or Safety Data Sheets (SDS) will specify what must be on the placards. National Fire Protection Association (NFPA) placards act as an immediate warning system for emergency service personnel, helping them identify the kinds of material present and the dangers they pose (Figures 5 and 6).

Benefits of Proper Storage

Storing pesticides poses hazards because multiple pesticide concentrates are kept together in a small area. However, a well-designed pesticide storage area will:

- Limit access,
- Allow better inventory control,
- Protect people from exposure,
- Reduce the chance of environmental contamination,
- Prevent damage to pesticides from temperature extremes and excess moisture,
- Safeguard pesticides from theft, vandalism, and unauthorized use, and
- Allow fire departments to know the location of products.

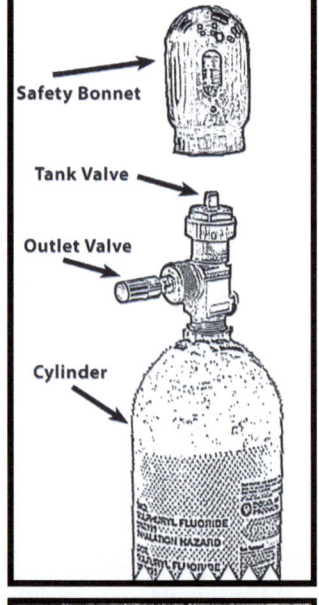

Figure 4: Storage of Fumigants. It is extremely important to secure fumigants so they will not move or get damaged during transport. Note that these cylinders are held fast with multiple ratchet straps (top left). Always make sure the safety bonnets are screwed on over the valves except when in use (top right). Fumigant cylinders may also come with covers over the outlet valves (bottom right) that also need to be in place until use. (Photo and illustration credits: (top left) © Garo Goodrow, Penn State; (top right) diagram of cylinder based on © photos from Garo Goodrow, Penn State; (bottom right) © Dale Dubberly)

STORAGE AND HANDLING: Store in a dry, cool, well-ventilated area under lock and key. Post as a pesticide storage area. Persons moving, handling, or opening containers must wear the personal protective equipment (including prescribed respirators when necessary) specified in the Hazards to Humans section of this labeling. Open container only in a well-ventilated area.

Figure 5: Storage and Handling Requirements on a fumigant label. A section from a methyl bromide product label showing some of the specific storage and handling requirements for that product.

Storage of Fumigants

In addition to the general requirements in California regulations, the fumigant label often has specific storage requirement instructions. Most fumigant labels direct users to store the product in a dry, cool, well ventilated, secured, and locked area that is posted as a pesticide storage area. For example, the ProFume® label recommends not to store the product indoors in occupied buildings. However, if the cylinders are found to be stored in an enclosed area without proper ventilation, the air in the area must be tested so persons entering or working in the space will not be exposed to sulfuryl fluoride concentrations greater than 1 ppm. Aluminum phosphide labeling may direct users not to store the product in areas where temperatures exceed 130°F and not to store the fumigant in buildings where humans or animals may reside.

Cylinder Storage

In general, always keep cylinders upright and secured so they will not fall over during storage or while you are using them. Keep the steel bonnet on the cylinder and remove it only when performing a fumigation operation. When not in use, remove the hose and be sure to screw the steel bonnet back onto the top of the cylinder. When a cylinder is empty, tightly close its valve by turning to the right, disconnect the hose, and replace the bonnet. Promptly return empty cylinders to the supplier. Never use cylinders for any other purpose or to store any other substance than it was intended for.

Never subject cylinders to rough handling or mechanical shock such as dropping, bumping, dragging, or sliding. Some cylinders have fusible plugs (heat sensitive plugs) that will rapidly release the gas if the cylinder reaches a certain temperature, so never expose cylinders to any heat source, including hot water. Do not store cylinders near flammable material, inlets of a ventilating or air conditioning unit, or any source of direct heat, or in a subsurface location.

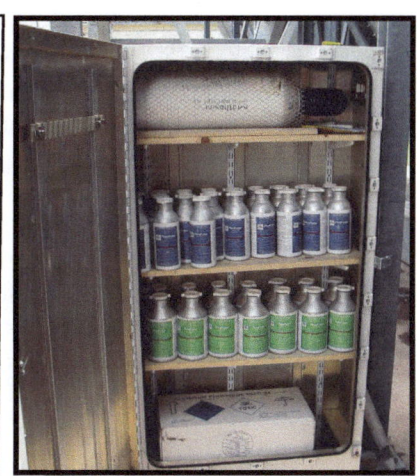

Figure 6: Pesticide Storage. Inside a fumigant storage locker (right). You must secure the storage area for any pesticide, but it's especially important for fumigants because of their high toxicity. California regulations require specific pesticide storage language (above). Important warning signs and placards are prescribed by state regulations and are also found on the label. (Photo credits: © J.Ples Spradley, pesticidepics.org at Virginia Tech).

```
DANGER
POISON STORAGE AREA
ALL UNAUTHORIZED PERSONS KEEP OUT
```

Inspect Valves and Containers

Fumigants can escape their containers through faulty, damaged, or corroded valves or containers. Leaks can cause dangerous fumigant concentrations to build up in closed areas. You should routinely check valves and containers for leaks using gas detection equipment. Before entering any storage area, run an exhaust fan to remove vapors that may have built up inside.

Other Storage Precautions

In addition to the information above, other precautions to take when storing fumigants include:
- Store cylinders secured in place to prevent them from falling over, with safety caps and protective bonnets securely in place.
- Wear any label-required PPE, including eye protection, when handling pressurized equipment.
- Always store fumigants (and other pesticides) in areas separate from food and feed.
- Never put fumigants in other containers.
- Keep metal fumigant containers off the ground to reduce moisture exposure, which can lead to rusting and leakage.
- Keep an updated and accurate inventory of all fumigants in storage.

Spills and Leaks

The fumigant labeling and Safety Data Sheet (SDS) will provide you with both general and detailed instructions on how to respond to pesticide spills, leaks, and fires. The label will tell you:
- When you need to wear a respirator,
- Whether the material can be salvaged, and
- What actions to take to minimize the risks to others.

If a spill, leak, or other sudden release occurs, evacuate the immediate work or storage area. Always put on appropriate PPE before attempting to move the leaking or damaged container outdoors or to an isolated area. Be familiar with and follow label instructions for handling leaks. Follow label requirements, such as using monitoring equipment, to determine the concentration of the fumigant in the ambient air. Generally, until the label threshold is met, you and anyone else in the work or storage area must wear an approved respirator if it is necessary to enter the area. If possible, move the leaking or damaged cylinder outdoors or to an isolated location. Work upwind if feasible. Get immediate medical attention for anyone who has inhaled a fumigant, even if they do not exhibit exposure symptoms.

Major Spills
The clean-up of a major spill may be difficult for an individual to do by themselves, or if an applicator is unsure what protocols to follow. Always follow state and/or local procedures. Contact the local agencies (fire department and police department) and the local CAC office for further assistance and information. Refer to the Safety Data Sheet (SDS) of the product that has been spilled for guidance on proper spill and clean-up procedures. Another useful resource is the Chemical Transportation Emergency Center (CHEMTREC) that can be reached at (800) 262-8200.

Emergency Response Plans
In California, employers are required to have an accident response plan at the fumigation worksite to provide instructions that protect employees during situations such as spills, fire, and leaks. At least two essential pieces of information should be included in the Accident Response Plan. The first should be information regarding the security of the area where the problem occurred, such as securing the area to prevent potential exposure until help arrives. The second is information on whom to contact in the event of a problem. Contacts may include the operator of the property, fire department, heath department, and/or hazardous materials response team.

Many fumigant labels, as part of creating a "Fumigant Management Plan" (FMP), will direct you to prepare a written Emergency Response Plan. This plan will usually contain explicit instructions for dealing with a spill or other emergency. It will list the names, and telephone numbers of people and local authorities to notify if fumigant levels in an area reach concentrations that could be dangerous to bystanders and/or domestic animals. FMPs will be covered in more detail in Chapter 4, "Fumigation Safety."

Safe Disposal of Leftover Fumigant
All labels indicate that "Improper disposal of excess pesticide is a violation of Federal Law." In addition to being required by law, it is common sense to properly dispose of highly toxic compounds, such as fumigants, so they do not cause harm to people or the environment.

You can contact your local County Agricultural Commissioner (CAC) or hazardous waste specialist for guidance on proper disposal.

The disposal of leftover fumigant is specific to the fumigant itself and the packaging of the product. For example, if you have some cylinders of methyl bromide or sulfuryl fluoride with unused product left, you should first contact the company before returning them.

Solid Phosphide Products
Leftover aluminum or magnesium phosphide products are handled differently. When properly exposed to moisture and expended, aluminum or magnesium phosphides will leave a grayish-white powder residue after a fumigation. This residue is nonhazardous waste. On the other hand, residual dust from incompletely exposed products called "green dust" must first be deactivated before disposal. The label will list the specific procedures for deactivation. Deactivation procedures may vary depending on the product formulation (i.e., pellets, tablets, gas bags).

Never store spent residual phosphide dust in a confined space or closed container. Confinement, collection, and storage of large quantities of dust may result in a fire hazard.

Management of Empty Containers
As always, the label is your best source of information on proper disposal of empty containers. Many fumigant products are contained in pressurized cylinders which are returned to the company for refilling. The labels will detail the process for the return of cylinders including how to indicate an issue. Other fumigants, like MITC and the MITC-generating fumigants, may be in steel or plastic containers.

Aluminum Flasks

Aluminum flasks that contained phosphides are nonrefillable containers. **Do not** reuse or refill aluminum flasks or use them for any other than their intended purpose.

Some labels will direct you to triple rinse empty flasks and stoppers with water. Afterwards, they may then be recycled or reconditioned or punctured and disposed of in a sanitary landfill, but not through household recycling.

For some products, the label allows for the removal of lids and exposing the empty flasks to atmospheric conditions until any incompletely exposed phosphide solids have reacted and only non-hazardous residue is left in the flask. In this case, puncture and dispose of the flask in a sanitary landfill or other approved site, or by other procedures approved by state and local authorities.

When a Fumigation Goes Wrong
Real World Example

What Happened?

A farmer found an old box of phosphine canisters in a storage room in his barn. He moved the canisters, along with other things he was discarding, outside to load onto his truck and take to the dump later in the week. It rained before he was able to transport the trash to the dump. The phosphine canisters were not well sealed (probably due to age) and the rain activated the tablets. The farmer had beehives nearby and all the bees died.

What Should Have Happened?

The farmer should have read the labels on the canisters or called the manufacturer of the product to find out about proper disposal of the canisters.

How It Could Have Been Avoided?

The farmer should have been aware that fumigants cannot be disposed of like other trash. He could have also read the pesticide label or called the manufacturer to get information on the proper disposal of the canisters.

What Was the Consequence?

Bees in the farmer's beehives died. The consequences could have been worse if people and/or non-target organisms were nearby and impacted.

Chapter 2 Review Questions

Correct answers are given on page 175.

1. Metal fumigant containers should be stored off the ground to _____.
    a. stabilize their pressure
    b. prevent cross-contamination
    c. reduce moisture exposure

2. Who is responsible for identifying sensitive sites on a restricted materials permit application?
    a. County Agricultural Commissioner
    b. Property Operator
    c. Certified Applicator

3. John plans to conduct commodity fumigations using a restricted material in Sacramento, Monterey, and Tulare county. How many restricted material permits does John need to apply for?
    a. one, from the county he resides in
    b. two, from the two largest counties
    c. three, one for each county

4. Which of the following statements is **NOT** true of the instructions in the applicator's manual?
    a. They are just suggestions.
    b. They are legally binding.
    c. They are part of the fumigant's labeling.

CHAPTER 3

Planning for a Fumigation

LEARNING OBJECTIVES

- ☑ Describe factors to consider when selecting an application method to use for a particular site.
- ☑ Explain the importance, and outline the process of, inspecting the application site prior to fumigation.
- ☑ List and explain how certain site characteristics and environmental factors can impact a fumigation.
- ☑ Explain why label statements may limit applications when conditions are unfavorable.
- ☑ Understand how to calculate the amount of product required for a treatment area.
- ☑ Describe how to determine area and volume of a structure to be fumigated.
- ☑ Explain the importance of sealing a fumigation site.
- ☑ List the methods used to seal an area for fumigation.
- ☑ List the two main reasons for air monitoring of fumigant levels.
- ☑ Explain when, where, and how to take air samples.
- ☑ Discuss why proper calibration of your monitoring equipment is important.
- ☑ Compare and contrast equipment you might use to monitor site fumigant levels.
- ☑ Explain when and how to notify authorities about a fumigation.

Terms to Know

The following are important terms to know from this chapter. They will be *italicized* in their first use. They are explained in the text and defined in the glossary at the end of this manual.

Air Monitoring	CT Concept	Metabolism	Tarpaulin / Tarp
Ambient Air Analyzers	Detectors	Sealing	Thermal Conductivity Analyzers
Dose	Detector Tubes	Stratification	
Dosage	Halide Leak Detectors	Tape-and-seal	

All methods of non-soil fumigation share a common requirement: the need to contain an adequate concentration of fumigant for the time necessary to kill pests. Non-soil fumigants may be applied in sites such as grain bins, bulk storage facilities, specially designed chambers, rail cars, trucks, ships, shipping containers, processing plants for commodity storage pests, tarped commodities in the field, rodent burrows, wine barrels, utility poles, and more. A correctly performed fumigation can solve pest problems.

Often your choice of fumigation method will be dictated by the site, type of pests, or commodity needing to be fumigated. Do you need to treat fresh produce? Packaged foods? Vehicles? Museum specimens? Furniture? Other small items? Once you've determined what will be fumigated, you would then likely choose between chamber or tarp fumigation. Spot fumigation, a short-term treatment, uses a tarp or tape-and-seal of machinery and may take place in small areas within a larger structure.

Premise Inspection

A pre-fumigation premise inspection is an important step in any fumigation process and is usually required by the Fumigation Management Plan (see Chapter 4, "Fumigation Safety"). Prior to a pre-fumigation premise inspection, you should have already determined that there is a pest problem that needs to be controlled and that a fumigation is the best control method.

A site inspection helps you determine what fumigant to use, how to conduct the fumigation, and allows you to assess any safety concerns and plan to address them. The success of the fumigation will depend on what you learned from the pre-fumigation premise inspection, your proper choice in control method for the pest problem, and how you plan for and conduct the fumigation.

Premise Inspection Items

The list below is not exhaustive but is meant to guide you as you familiarize yourself with the fumigation site during a premise inspection. Not all items apply in all situations. For example, what may apply in a commodity fumigation may not apply in a wine barrel fumigation or in sewer line root control. Always follow the instructions on the product label and in the applicator's manual.

- Determine whether the structure itself or just certain items in the structure are infested. If the latter, it may be easier to remove infested items, if possible, and fumigate them elsewhere (e.g., in a fumigation chamber) or fumigate the items under a tarp so that you don't need to treat the whole structure. Be aware that it may not be possible to remove or move items that have been quarantined by a government agency.
- Inspect the structure or area to determine its suitability for fumigation. What is the structure made of? Can you achieve a tight seal with a tape-and-seal method or should you use a tarp? If you will be tarping material inside the structure, does the structure have a smooth, impermeable floor to establish a tight chamber?
- Determine if the fumigant you will use will require removing or protecting certain materials or living things before fumigation.
- Know the location of the pests. For example, are they inside the commodity or on its packaging?
- Locate all places that need to be sealed. For example, if you are fumigating a structure housing a commodity, check for broken or cracked windows, ceilings, walls, or floors. Also look for floor drains, sewer pipes, or cable conduits that could allow the migration of fumigant into another area or structure. Air ducts and ventilation fans are other potential spots where the fumigant can escape or leak.
- Check for connected structures. If there are parts of the building not under the control of your client, determine if these parts can be shut down and evacuated during the fumigation.
- Determine gas and electrical concerns. For example, the ProFume® label requires you to locate and turn off all gas cutoffs, pilot lights, and electric heating units in the structure. All must be off during the fumigation to prevent fire and/or adverse reaction of the fumigant to heat. For many fumigants, knowing the location of electrical outlets will help you decide where to place circulating fans.
- Plan for aeration. Determine how to aerate the fumigated space. Be familiar with the location of any exhaust fans and their switches. Find out if neighboring buildings have air intakes which could draw in the fumigant during aeration. Consider and take precautions to protect neighboring sensitive sites, such as nearby schools or residences, that may be affected during aeration. Neighboring buildings may have air intakes that could draw in fumigant during aeration—you must ensure that neighboring property, as well as people, are not exposed to the fumigant.
- Do not ignore the outside of the fumigated space. Be certain that you can create a tight ground seal if a tarp is needed. Be aware of other employees or bystanders who may be near the fumigation site.

Site Characteristics and Environmental Conditions that Affect Fumigation

There are a number of factors that you need to consider to ensure that a fumigation will be both safe and effective. Following are some important factors that affect a fumigant's performance.

Type of Fumigation

For fumigations in a structure (e.g., a fumigation chamber or grain bin), the type of structure, what it is made of, and how airtight it is (or can be made with sealing) can influence the success of the fumigation. Wooden structures, even when tightly constructed and well-sealed, do not retain fumigant as well as steel, masonry, or concrete structures. In general, concrete structures retain fumigant better. Similarly, round steel bins retain fumigant better than flat grain storages.

When a temporary chamber is created using tarps or other impermeable sheeting, these tarped stack fumigations rely on the quality of the tarps or sheeting, as well as the seal, to retain the fumigant. The items being fumigated should be placed on an airtight surface, such as a tarp or concrete. If the commodity being fumigated is on wood or other porous material, some labels may instruct you to move the commodity onto a nonporous material such as a tarp.

Temperature

Temperature at the treatment site affects both the fumigant itself and the target pest.
- **Effects on Fumigation:** Temperature affects the release rate and the speed of penetration of the fumigant. As temperature increases, the fumigant becomes more volatile, is released more rapidly, and disperses and penetrates more quickly than at lower temperatures. Also, both the *dosage* and exposure time vary with temperature differences. Temperature differences can cause a fumigant to "stratify." *Stratification* occurs when the air and fumigant form layers and do not mix. In general, stratification is more severe when the temperature of a fumigant is significantly lower than that of the air.
- **Effects on Target Pests:** In general, insect pests are more difficult to control at lower temperatures. At temperatures below 50°F, many stored commodity insect pests are dormant and thus, exhibit little or no activity. In these conditions, insects breathe at a very minimal rate and they may not take in enough fumigant to kill them—or the exposure time needed to kill them would have to be longer than allowed by the label. Insect activity (respiration, feeding, and growing) will differ by species but is generally more rapid as the temperature increases up to a maximum. Although many label requirements state a minimum of 40°F to fumigate, fumigating at temperatures of 60°F or higher is generally more effective for insect pests. Increasing the temperature (up to a certain point) increases the *metabolism* of all insect life stages causing them to take in more fumigant because they respire more rapidly. This means that less fumigant (or less time) is needed to provide effective control. Preferred fumigation temperatures usually range between 50°F and 95°F, though some fumigant labels allow for cold storage fumigations down to 28°F. <u>Always</u> consult the label for the fumigant's acceptable temperature range and optimum temperature.

 **NOTE:** There are usually no low temperature restrictions on fumigants used for rodent control.

Humidity

Humidity can affect the performance of water-soluble fumigants such as methyl bromide. Water soluble fumigants become unavailable when dissolved in water, reducing their concentration. Fumigants may not be able to penetrate wet or damp areas, allowing insect pests in those areas to survive. High humidity can lead to condensation which can cause wet spots that allow molds or other damaging conditions to develop. As the moisture content of a commodity increases, it becomes more difficult for a fumigant to penetrate the commodity and reach the target insect pests. The increased moisture of a commodity may also result in an increase of pesticide residues left on the commodity, which may exceed legal tolerances.

On the other hand, aluminum and magnesium phosphide require adequate moisture to generate phosphine. If the air is too dry or the moisture content of the commodity is too low, these fumigants will stay in solid form.

Sorption

The sorptive quality (adsorption or absorption) of the treated commodity and/or the structure can affect how long it takes a fumigant to reach the desired concentration for effective pest control. Sorption also affects the duration of the aeration process.

Air Movement

Do not fumigate on a windy day. Winds can result in considerable loss of fumigant even in a well-sealed structure. For example, winds around a grain storage structure create pressure gradients across the grain surface which can cause rapid loss of fumigant.

In *tarpaulin* (tarp) fumigation, winds cause tarps to billow and separate from what they cover. Excessive billowing can speed the loss of fumigant under the tarp. You can prevent this by keeping the tarp tight against the structure or fumigated commodity. Because of added stress on tarps due to wind and weight, you should decrease the clamping intervals (i.e., put clamps closer together) and increase the weight on tarp aprons (the section of tarp on the ground). You might also need to use ropes to secure tarps.

However, within the commodity or space being treated, some air movement is essential for effective fumigation. The gas must spread evenly and quickly to enter small crevices, cracks, or spaces so that a lethal concentration contacts every pest. Fumigants also spread faster when their initial concentration is high and the penetration distance is short. Fans are useful and may be necessary to uniformly distribute fumigant so that it reaches all target pests. Fans may also be necessary to ensure penetration of fumigant into bulk commodities. When conducting a fumigation with methyl bromide or sulfuryl fluoride, you must use non-sparking fans and locate and turn off heat sources (e.g., pilot lights, electric heating units, etc.)—heat sources cause methyl bromide and sulfuryl fluoride to become corrosive and can cause phosphine to ignite.

 The product label and applicator's manual may prohibit use of the fumigant when factors, including those discussed above (or others), make it unfavorable to conduct a fumigation. Unfavorable conditions could result in the failure of the fumigation to control the pest and/or adversely expose non-target animals, people, or the environment to the fumigant. As always, read and follow label directions.

Calculating Fumigant Dosage

Understanding the relationship of fumigant concentration, exposure time, and temperature during fumigation is critical in determining the correct dosage to control the pest of interest. For sulfuryl fluoride and methyl bromide, the "*CT (Concentration X Time) concept*" is used to calculate the dosage of the fumigant.

Dosage (D) = Concentration (C) × Time (T)
or, in short:
D = C × T

We use a few abbreviations in the following section. You may already be familiar with them, but we list them below for clarification and convenience.		
ft = feet	ft² = square feet	ft³ = cubic feet
l = length	w = width	h = height
r = radius	r² = radius squared (or r x r)	

You can achieve the needed fumigant dosage by changing either C or T to produce a toxic effect on the target pest. Increasing the exposure time (T), decreases concentration of the fumigant (C) required to achieve the dosage (D) for control. Conversely, decreasing the exposure time (T), requires increasing the concentration of the fumigant (C) to get to the appropriate dosage (D) for control.

 NOTE: The formula above does not apply to phosphine because the relationship of dosage and toxicity of phosphine to insects is not linear. Phosphine is most effective when exposure times are 1 day or longer. In general, longer exposures to phosphine at lower concentrations result in better efficacy than shorter exposures at higher concentrations.

All fumigant labels will list recommended dosages to effectively treat the target pest. Never exceed the highest dosage on the label. Some manufacturers supply special calculators or computer-based programs to determine the dosage of their products.

Calculating Area and Volume

The recommended rates, or *doses,* listed on fumigant labels are based on the volume of the space to be fumigated. It is critical to accurately determine volume of the treatment area before you begin a fumigation. If you use too little fumigant for a given volume, you may not achieve the desired level of pest control. Using too much fumigant is wasteful, can damage the treated commodity, and is illegal if concentrations are above the label rate.

Determining Volume

The volume (or cubic content) of a structure is equal to the structure's area multiplied by its height. The area is determined by multiplying the length by the width. As below:

- Area = length (l) × width (w)
- Volume = length (l) × width (w) × height (h)

Length, width, and height are measured in feet, area in square feet (ft^2), and volume in cubic feet (ft^3).

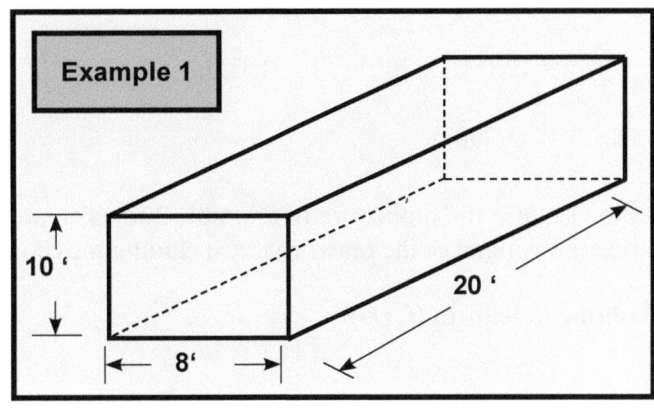

In this chapter, we will show you how to calculate the volumes of differently shaped structures. Each example will be slightly more complicated than the one before it.

The structure in **Example 1** is a fumigation chamber 20 feet long (l), 8 feet wide (w), and 10 feet high (h). What is its volume?

$$\begin{aligned}\text{Volume} &= \text{length} \times \text{width} \times \text{height} \\ &= 20 \text{ ft} \times 8 \text{ ft} \times 10 \text{ ft} \\ &= 1{,}600 \text{ ft}^3\end{aligned}$$

In practice, calculating volume is usually more involved because many fumigation sites are irregular in shape.

For **Example 2**, the height of the fumigation chamber is different at different points along the roof. To calculate the volume of this structure you must first determine the average height. The easiest way to do this is to take the average of the wall height and the height at the peak. The average height of the structure in Example 2 is:

$$\text{Average Height} = \frac{\text{Wall height} + \text{Peak Height}}{2}$$

$$= \frac{10 \text{ ft} + 18 \text{ ft}}{2}$$

$$= 14 \text{ ft}$$

Thus: Volume = length × width × average height

$$= 40 \text{ ft} \times 30 \text{ ft} \times 14 \text{ ft}$$

$$= 16{,}800 \text{ ft}^3$$

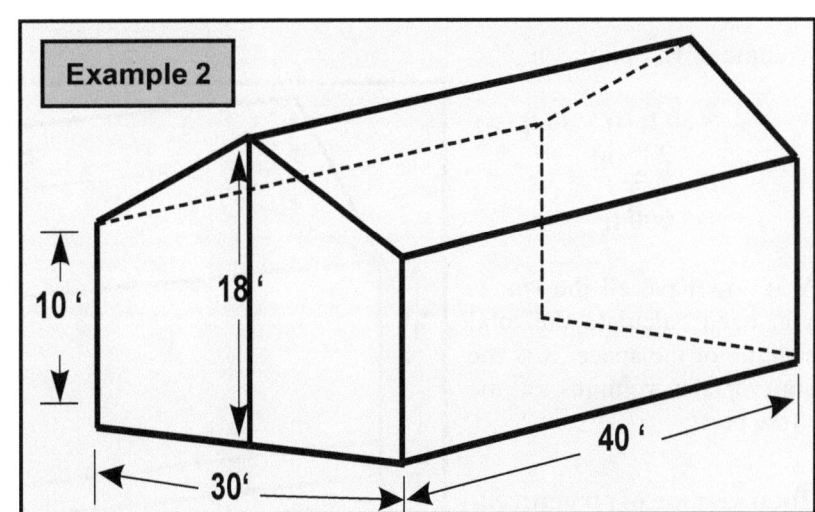

Let's now look at a more complicated space in Example 3.

There are two ways to compute the volume of the structure in **Example 3**. The first method is to partition the structure into three parts: the main room (A) and its roof (B), the side lean-to (C) and its roof (D), and the crawl space of the entire structure (E).

Use the procedure in Example 2 to first determine the average height of the main room. Use this average height to determine the volume of the main room, including the roof.

Volume of main room (A+B):

If the average height of the main room **(A)** + the roof in the main room **(B)** = $\frac{8 \text{ ft} + 14 \text{ ft}}{2} = \frac{22 \text{ ft}}{2} = 11$ ft

Then the volume of A + B = l x w x h

= 40 ft (l) x 30 ft (w) x 11 ft (h)

= 13,200 ft³

Likewise, use the procedure in Example 2 to calculate the volume of the lean-to. Use the average height to determine the volume of the crawl space, including the roof.

Volume of lean-to (C+D):

If the average height of lean-to crawl space **(C)** + the roof in the lean-to crawl space **(D)** = $\frac{6 \text{ ft} + 8 \text{ ft}}{2} = 7$ ft

Then the volume of C + D = l x w x h

= 30 ft (l) x 10 ft (w) x 7 ft (h)

= 2,100 ft³

Calculate the volume of the crawl space.

Volume of crawl space:

Volume of **E** = l x w x h

= 50 ft (l) × 30 ft (w) × 2 ft (h)

= 3,000 ft³

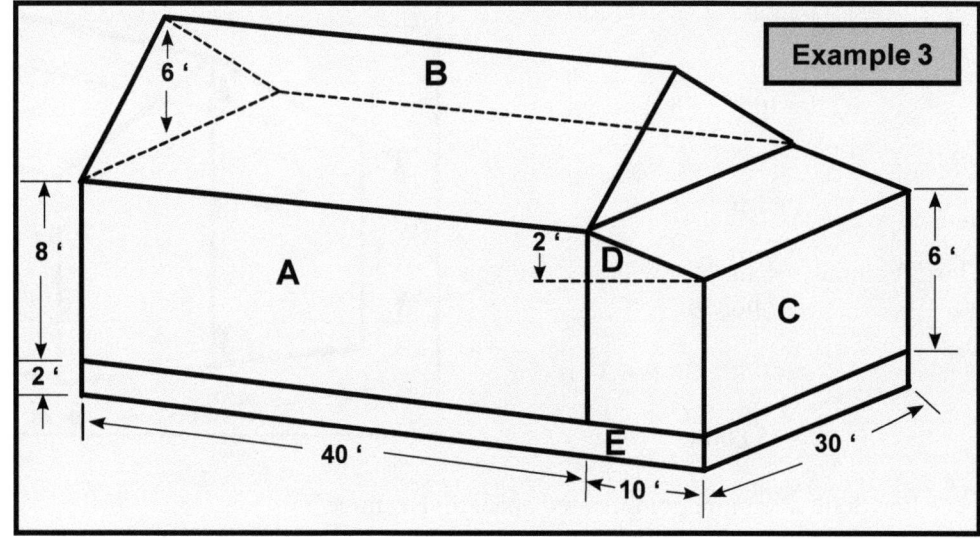

You now have all the values you need calculate the total volume of the space. It is the sum of the volumes of the three parts:

Total volume of structure:

Total volume = Volume of main room and roof + Volume of lean-to + Volume of crawl space

= 13,200 ft³ + 2,100 ft³ + 3,000 ft³

= 18,300 ft³

The second way to calculate the volume of the space in **Example 3** is to calculate the volumes of all five "parts" separately (the main room (A), the roof in the main room (B), the lean-to (C), the roof in the lean-to (D), and the crawl space (E)) then add these values to get the total volume.

This second approach is provided to show:
1. How to calculate the volumes of B and D; and
2. That you can obtain the total volume using different methods.

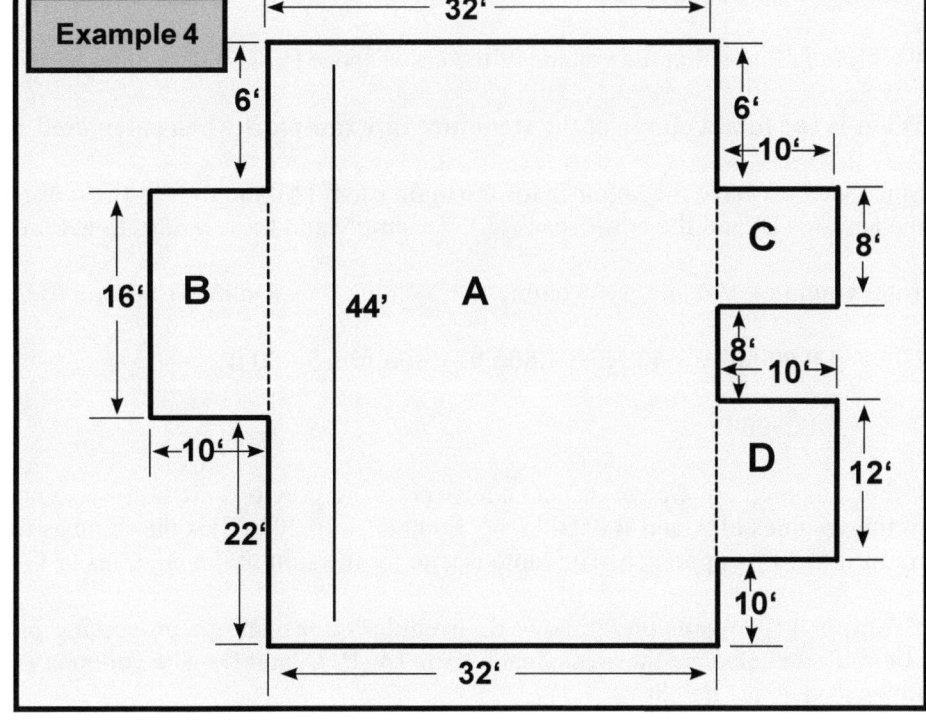

Example 4

Volume of the main room (A)

We already know how to calculate the volume of A (the main) room.

Volume of **A**: 40 ft (l) × 30 ft (w) × 8 ft (h) = 9,600 ft^3

Volume of the roof in the main room (B)

To calculate the volume of **B** (the roof) we need to determine the average height first. Since the roof does not have a "wall," the wall height = 0.

Therefore the average height of the roof (**B**) =

$$\frac{0 \text{ ft} + 6 \text{ ft}}{2} = 3 \text{ ft}$$

And the volume of roof (B) = l x w x h

= 40 ft (l) x 30 ft (w) x 3 ft (h)

= 3,600 ft^3

Volume of lean-to (C)

We already know how to calculate the volume of C (the lean-to).

Volume of C = 10 ft (l) x 30 ft (w) x 6 ft (h)

= 1,800 ft^3

Volume of the roof in the lean-to (D)

We can calculate the volume of the lean-to roof (D) the same way we calculated the volume of (B). Since the average height of lean-to (**D**) = $\frac{0 \text{ ft} + 2 \text{ ft}}{2} = 1$ ft

Then the volume of lean-to (**D**) = l x w x h

= 10 ft (l) x 30 ft (w) x 1 ft (h)

= 300 ft^3

Volume of crawl space (E)

We already know that the volume of the crawl space (E) is 3,000 cubic feet.

What is the total volume of the structure in Example 3 when calculated using the second method?

Since we now have the volumes for the main room (A), the roof in the main room (B), the lean-to (C), the roof in the lean-to (D) and the crawl space (E), we simply add these values to get the total volume.

Total volume = Volume A + Volume B + Volume C + Volume D + Volume E

$$= 9,600 \text{ ft}^3 + 3,600 \text{ ft}^3 + 1,800 \text{ ft}^3 + 300 \text{ ft}^3 + 3,000 \text{ ft}^3$$

$$= 18,300 \text{ ft}^3$$

Using this method, we get the same answer as the previous method. Also note that, using this approach, the sum of the volumes of **A** and **B** (9,600 ft^3 + 3,600 ft^3 = 13,200 ft^3) is the same as the volume we calculated for the main room in the first approach. The same is true for the sum of the volumes of **C** and **D**.

Example 4 (previous page) shows an irregular floor plan of a processing plant. The first step in determining the volume is to calculate the area of each room (**A, B, C,** and **D**). The volume of each room is equal to the room's area times its average height.

Volume **A** = Area **A** × Height **A**. Where Area **A** = Length **A** x Width **A**.

The total volume of the structure is the sum of the volumes of all four rooms:

Total Volume = Volume **A** + Volume **B** + Volume **C** + Volume **D**
= (Area **A** × Height **A**) + (Area **B** × Height **B**) + (Area **C** × Height **C**) + (Area **D** × Height **D**)

This equation is applicable even if the heights of the rooms differ, but calculations are much simpler if the height of the roof is the same in all rooms. Let's try calculating the volume of the space in Example 4. Assume that the structure in **Example 4** is covered with a flat roof and the height of each room is 10 feet (h). Calculate the total volume by adding the volumes of the four rooms:

Volume **A** = (Area **A** x Height **A**)
= (44 ft × 32 ft) × 10 ft
= 14,080 ft^3

Volume **B** = (Area **B** x Height **B**)
= (16 ft × 10 ft) × 10 ft
= 1,600 ft^3

Volume **C** = (Area **C** x Height **C**)
= (8 ft × 10 ft) × 10 ft
= 800 ft^3

Volume **D** = (Area **D** x Height **D**)
= (12 ft × 10 ft) × 10 ft
= 1,200 ft^3

The total volume of the space is:

$$\text{Total Volume} = \text{Volume A} + \text{Volume B} + \text{Volume C} + \text{Volume D}$$

$$= 14{,}080 \text{ ft}^3 + 1{,}600 \text{ ft}^3 + 800 \text{ ft}^3 + 1{,}200 \text{ ft}^3$$

$$= 17{,}680 \text{ ft}^3$$

Since the height of all the rooms in Example 4 are assumed to be the same (10 ft), calculating the volume of the space is simple. The total volume of the structure can also be calculated by multiplying the same height measurement by the sum of the areas of the four rooms:

Total Volume = Height × (Area **A** + Area **B** + Area **C** + Area **D**) = Height x Total Area

If Total Area = Area **A** + Area **B** + Area **C** + Area **D**
= (44 ft × 32 ft) + (16 ft × 10 ft) + (8 ft × 10 ft) + (12 ft × 10 ft)

= 1,408 ft² + 160 ft² + 80 ft² + 120 ft²

= 1,768 ft²

Then Total Volume = Height x Total Area

= 1,768 ft² × 10 ft

= 17,680 ft³

This value is the same as the volume we calculated earlier.

Remember, if the height of the structure varies in different areas, use the first method (Total Volume = Sum of Individual Volumes). Only use the second method (Total Volume = Height × Total Area) when the height is the same for all the rooms.

The examples we have looked at so far represent fairly simple structures. If a fumigated space is covered by a tarp, you must also account for any space under the tarp when calculating the total volume to be fumigated. While the task of calculating the volume may take longer in some cases, the basic methods of calculation are the same as in the examples here.

Round or Cylindrical Structures
Now let's look at a structure that is not rectangular. If you wish to fumigate an empty grain bin, how do you determine the volume of the bin?

Example 5 is a simplified illustration of a grain bin. Grain bins are a cylinder (like a big tin can) with a conical (funnel-shaped) cap. The total volume of the grain bin in Example 5 is calculated by adding the volume of the cylinder and the volume of the cone.

The formula for calculating the volume of a cylinder is:

Volume of Cylinder = π x radius² (r²) x height (h) = 3.14 × r² × h

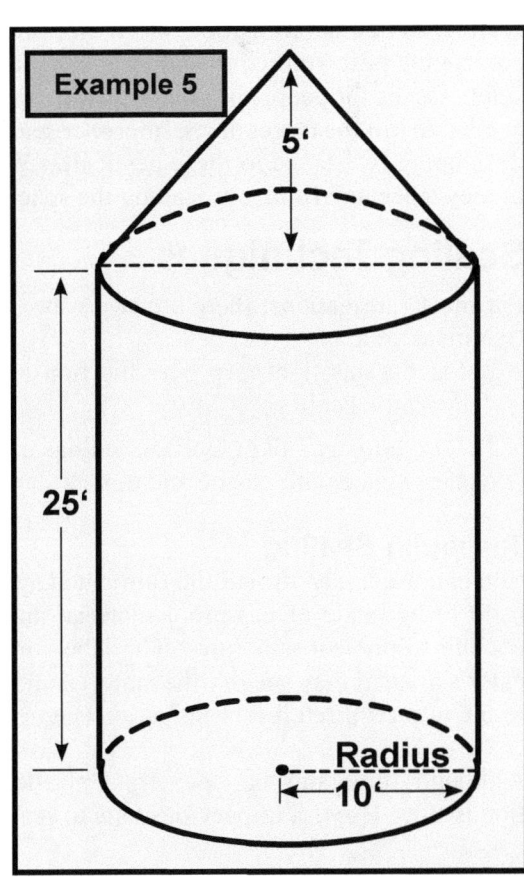

Example 5

The formula for calculating the volume of a cone is:

$$\text{Volume of Cone} = \frac{\pi \times \text{radius}^2 (r^2) \times \text{height (h)}}{3} = \frac{3.14 \times r^2 \times h}{3}$$

In both calculations, the Greek letter Pi (π) is the ratio of the circumference of a circle to its diameter and has a value of 3.14. The number 3.14 is a "constant"—the same value is used in the formula regardless of the values of r and h.

The radius is most easily measured at the base of the cylinder or cone and is equal to the distance from the center of the base to the edge. Note that in both formulas, the radius is squared (r^2) or (r x r), meaning the radius is multiplied by the radius.

Using the formulas above we can calculate the total volume of the grain bin in **Example 5**.

Volume of Cylinder = π x radius^2 (r^2) x height (h)

$$= 3.14 \times (10 \text{ ft})^2 \times 25 \text{ ft}$$

$$= 7{,}850 \text{ ft}^3$$

Volume of Cone = $\dfrac{3.14 \times (10 \text{ ft})^2 \times 5 \text{ ft}}{3}$ = 523.3 ft^3

Total Volume = Volume$_{\text{cylinder}}$ + Volume$_{\text{cone}}$ = 7,850 ft^3 + 523.3 ft^3 = 8,373.3 ft^3

Sealing Fumigation Sites

In addition to following the product label for temperature and dosage, another important factor to ensure the success of a fumigation is properly *sealing* the fumigated space. Almost all fumigation failures are due to inadequate sealing. When a fumigation site is not properly sealed, you cannot maintain the gas concentrations needed to control the pest. Improper sealing results in the fumigant leaving the treated space, which can lead to health and safety issues for people or animals exposed to it, and/or environmental concerns. Poor seals can lengthen the time needed to kill the target pests. Improper sealing, resulting in lower fumigant concentrations, can contribute to pests developing resistance to these pesticides. When the space is not properly sealed, it is also a waste of product and money since any fumigant leaving the space will not control the pests as intended.

Sealing Techniques

For most fumigations, there are two ways to seal the treatment area:

1. Place a gas tight tarp over the fumigated space (tarpaulin sealing).
2. *Tape-and-seal* all potential openings of the fumigated space with plastic and tape.

Tarpaulin Sealing

You can use a tarp to seal the fumigated space. Tarps need to be made of gas-proof material and be thick enough to prevent punctures. The label and/or applicator's manual may specify the material and thickness of the tarp required for a fumigation (Figure 1).

Although fumigants can penetrate plastic, penetration is slow. Use gas impervious tape to seal all seams

---

Tarp Material

Plastic tarps are semi-permeable membranes that permit different fumigants to pass through them at different rates. The passage of ProFume through most plastic sheeting of sufficient thickness is very slow (see Table 5a).

Use only tarps made of materials that will adequately confine ProFume for the required time. Tarps are sold in many sizes. Experience has shown that the following have proven satisfactory:

1. 4 to 6 mil polyethylene for "single use" tarps
2. Laminated (several layers) polyethylene
3. Vinyl coated nylon
4. Neoprene coated nylon
5. PVC (polyvinyl chloride) coated nylon

Thickness

As a minimum, 4 to 6 mil (160 to 240 microns) thickness of the above materials adequately confines ProFume. A tarp of 100 microns is equivalent to a 400-gauge material. Polyethylene tarps less than 4 mil (160 microns) are not of an adequate thickness to confine ProFume because they do

Figure 1: Tarp specifications found on an applicator's manual. Taken from a ProFume® applicator's manual, this section specifies the tarp requirements. The applicator's manual and/or the label will also include information and guidance on sealing.

where tarp or plastic join. Good sealing is more important than the thickness of the tarp or tape.

You must seal the tarp to the ground with loose wet sand, snakes (flexible tubes filled with water or sand), adhesives, or a combination of these methods to prevent gas from leaking out. If using snakes, use two rows of snakes along the sides and three rows on the corners. Overlap the snakes by about one foot. You can use loose wet sand to seal points where fumigation equipment goes under the tarpaulin (e.g., fumigant introduction line, electrical cords, gas sampling tubes).

Tape-and-Seal
Structures built with impermeable exteriors such as concrete or steel are most suitable for tape-and-seal fumigation. For tape-and-seal fumigation, use spray adhesive and tape to affix plastic sheeting to doorways, vents, and other openings. In grain bins, you should also seal unloading augers, roof exhaust vents, and eave gaps (openings where the roof meets the sidewalls).

Follow these guidelines when sealing:
- Make sure you have a clean, firm, and dry surface to seal to.
- For grain bins, seal outside whenever possible to minimize the need to enter the confined space containing grain.
- If you anticipate windy conditions, you might need additional sealing.

> **C. MONITORING**
> 1. Safety
>   a. Monitoring of phosphine conditions must be conducted in areas to prevent excessive exposure and to determine where exposure may occur. Document where monitoring will occur.
>   b. Keep a log or manual of monitoring records for each fumigation site. This log must at a minimum contain the timing, number of readings taken and level of concentrations found in each location.
>   c. When monitoring, document even if there is no phosphine present above the safe levels. In such cases, subsequent monitoring is not routinely required. However spot checks must be made occasionally, especially if conditions significantly change.
> 2. Efficacy
>   a. For stationary structures, phosphine readings MUST be taken from within the fumigated structure to insure proper gas concentrations. If the phosphine levels have fallen below the targeted level, the fumigators, following proper entry procedures may reenter the structure and add additional product.

Figure 2: Monitoring information on an applicator manual. This section was taken from a phosphine fumigant product applicator's manual. Monitoring is required for both the safety of workers and the efficacy of the application. Applicators must keep a log of all measurements.

Other Techniques
Other fumigation activities may require specific methods to seal the fumigated space. For example, when fumigating wine barrels with SO_2, the barrel opening is immediately sealed after removing the SO_2 probe. Wine barrel corks are fumigated in a plastic bag that is heat-sealed. For utility pole fumigation, the hole drilled to introduce the fumigant is sealed with a tight-fitting wooden dowel or cork to keep the fumigant in the pole. For rodent burrow fumigation, you often seal the burrow with soil or turf to keep the fumigant in the space.

Permanent Sealing
For any fumigation site, especially one you know you will treat more than once, your job will be easier by permanently sealing cracks and unused openings. You should always get prior approval from owners or managers before doing this. You can use sealants to cover cracks and unused openings around doors, and equipment, or electrical conduits through walls, ceilings, or junctions that penetrate walls to the outside or to non-fumigated areas.

In bins, you can permanently seal:
- Exterior under-roof vents.
- Roof deck-to-wall gaps.
- Clearance openings around centrifugal direct drive motor shafts.
- Bolt holes with missing bolts.
- Gaps between flanges on aeration fans connected to transition ducts.
- Aeration duct entrances through the bin wall or concrete foundation at the base.

Fumigant Monitoring

Air monitoring of fumigant concentrations is an extremely important (and sometimes mandatory) part of the fumigation process. Air monitoring involves the use of sensitive gas monitoring devices during fumigation to accurately gauge fumigant levels in relation to the dosage on the label and/or to detect leaks from the application site. Fumigant labels may require the use of sensitive gas monitoring devices during the application and before warning signs can be removed after the end of the application (Figure 2). Gas concentration is monitored during fumigation to make sure that an adequate concentration is maintained long enough to be effective. It is just as important to know the gas concentrations that you and other workers may be exposed to so that proper precautions, such as using respiratory equipment, can be taken.

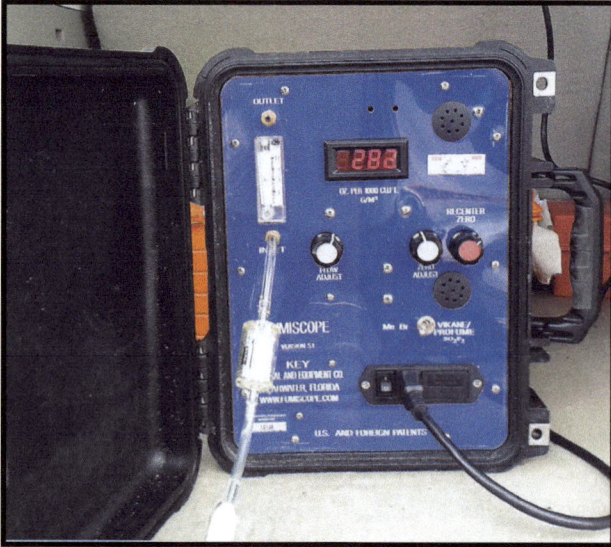

Figure 3: A fumiscope. A fumiscope is generally used to measure levels of methyl bromide or sulfuryl fluoride. Monitoring for the effective levels of a fumigant is important to ensure that the fumigation process is effective and successful. (Photo © Carl Schnabel).

 For safety reasons, it is essential to have an accurate and immediately available way to determine fumigant concentrations. Some fumigants have no detectable odor. Even fumigants with strong odors may not be detected by a worker with a poor sense of smell. Sensitive air monitoring equipment can accurately measure levels of fumigants that may not have strong odors or that are not detected by some people.

You may need to monitor both inside and outside the space being fumigated during and after the fumigation. When measuring fumigant levels, it's important to take readings from several locations because fumigants may become trapped in local pockets. Also, different materials and commodities will desorb at varying rates, a process called "off gassing," which can allow toxic levels of the fumigant to occur in scattered locations. The two main reasons you need to perform air monitoring during a fumigation are to ensure:
1. The efficacy of the fumigation, and
2. The safety of applicators, other personnel, and bystanders. On labels, this aspect of monitoring is often referred to as "industrial hygiene monitoring."

Monitoring for Efficacy

If the fumigant concentration does not reach the desired level because of a leak or insufficient pesticide rates, you can expect ineffective pest control possibly leading to a costly secondary application or uncontrolled pest damage. For most fumigations, you can use high concentration monitoring devices to check that the effective fumigant concentration specified on the label has been reached. Collect air samples through lines (typically 1/4 inch tubing) set up inside the fumigated enclosure prior to introducing the gas. These lines extend to a spot outside the enclosure and connect to a *detector* that takes periodic readings (or measurements) (Figure 3).

Measuring fumigant concentrations inside structures, chambers, or tarped products allows you to:
- Confirm that the correct concentration of gas has been introduced into the site.
- Determine when the gas inside the site is evenly distributed and when to start timing the exposure period.
- Make dosage adjustments if needed.
- Determine if, and when, the target endpoint (accumulated dosage of gas concentration x time) has been reached.

Monitoring for Safety

For most fumigations, the fumigant levels must be monitored periodically during the fumigation process in areas where applicators and bystanders may be exposed to the pesticide. Monitoring is essential to identify leaks that could be harmful to people (or bystanders). Product labels will instruct you on how and where to conduct monitoring for safety. Generally, low concentration detectors are used for safety monitoring (also referred to as "industrial hygiene monitoring").

After the gas has been introduced to a fumigated space, use a low concentration gas detector to check for fumigant leaks around the exterior perimeter. Use the detector to check seals and to ensure that gas is not lost from leaks. Repair leaks to minimize loss of fumigant and to reduce the risk of exposure to bystanders and/or occupants of nearby buildings. Table 3-1 summarizes the levels of three fumigants outside an enclosure that require leak repair.

Table 3-1. Fumigant concentration outside enclosure requiring leak repair

Fumigant	Gas Concentration (PEL*)
Phosphine	> 0.3 ppm
Sulfuryl Fluoride	> 1.0 ppm
Methyl Bromide	> 5.0 ppm

* PEL = Permissible Exposure Limits. This is the maximum amount or concentration of a chemical that a worker may be exposed to under OSHA regulations. Usually expressed in parts per million.

After fumigation is complete and the structure has been aerated, air monitoring is necessary to ensure a safe re-entry. Never enter an enclosed fumigated area, even after venting, without first measuring for toxic levels of fumigant vapors. Depending on the type of monitoring device you use, you may need to wear PPE (required respirators and protective clothing) when taking air samples. Some monitoring equipment allows for remote sensing. Take measurements in several locations within a confined space since localized pockets in the area sometimes trap vapors.

Some pesticide labels require applicators to keep a log to record gas concentrations below threshold levels for any vulnerable sites identified in the site-assessment (Figure 4).

Gas Detection Devices

There are a wide variety of gas detecting devices that can sample air for fumigants. Choose air-monitoring equipment designed to measure the levels of the fumigant you are using. All detection devices must be used and maintained according to the manufacturer's instructions and recommendations. Many types of detection devices require periodic calibration to ensure accurate measurements. Different equipment is needed depending on the need for low or high concentration monitoring. Prior to purchasing any equipment, read the pesticide labels of all products that will be used to determine the monitoring equipment needed and if there is a specific sensitivity required. The label and/or applicator's manual may prescribe specific detection equipment you need to use.

Following are some of the more commonly used gas detection devices in fumigation at the time this study manual was prepared.

Ambient Air Analyzers

Some *ambient air analyzers* (also called "infrared detection systems" or "IR devices") use infrared light to detect and measure gas concentrations (Figure 5). When infrared radiation strikes a gas, certain wavelengths of the radiation are absorbed. The device then measures this absorption to determine the gas concentration. Most ambient air analyzers can be calibrated at the factory to detect a single gas. Others are equipped with a fixed infrared filter and allow detection of multiple gases.

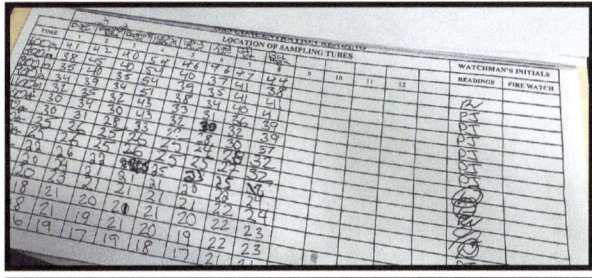

Figure 4: Regardless of requirements, it is good fumigation practice, and protection from liability, to keep records showing that all such thresholds have been met. (Photo © Ed Crow, Penn State).

Halide Leak Detectors

Halide leak detectors are useful for detecting the presence, and approximating the concentration, of methyl bromide and sulfuryl fluoride. These detectors are used to detect leaks and to determine if the fumigant is present in a work area but may not be appropriate for detecting potentially harmful concentrations. The exposure limit to ensure the safety of people is below the detection limit of these devices. Gas detector tubes that measure low concentrations of fumigants should be used to ensure worker safety because halide detectors are not intended to measure exact concentrations.

Be aware that halide detectors have an open flame when in use. Even when the detector is not in operation, do not store it in a frequently inhabited room. The fuel is a flammable gas under pressure and may explode. Do not use halide detectors in the presence of flammable or explosive gases such as gasoline vapors. Do not use halide detectors in mills, grain elevators, or other enclosures where there is a possibility of a dust explosion.

Thermal Conductivity Analyzers

Thermal conductivity analyzers (TCAs) measure the concentration of fumigant gases within a chamber or other enclosure during fumigation. Several types of TCAs are available. The Fumiscope® is one of the more common TCAs. It can measure methyl bromide and sulfuryl fluoride concentrations.

Figure 5: Monitoring with an infrared gas monitor. The worker is taking a reading outside a warehouse door for sulfuryl fluoride using an infrared gas monitor. (Photo © Garo Goodrow, Penn State).

 TCAs cannot measure gas concentrations below 5 ppm and, like a halide detector, should NOT be used to determine whether fumigant levels are safe for reentry.

Detector Tubes

Detector tubes (also called colorimetric tubes) are still being used but not as much as in the past. They can measure very low concentrations of a fumigant and are used to detect leaks and to determine if entering a fumigated space requires the use of respiratory protection. Detector tubes are sealed glass tubes filled with an indicator chemical that reacts with a specific fumigant to produce a color change (Figure 6). A measured amount of air is drawn through

the tube by a pump for testing. The amount of color change in the tube indicates the gas concentration in the sampled air. Extension tubes are available for sampling hard-to-reach areas. These detectors are easy to use and are reasonably accurate when used according to directions.

In addition to tubes that give immediate reactions, detector tubes can also be designed to monitor gases over a period of time, such as a work day or work shift. The tube is connected to a lightweight pump that continuously draws a measured volume of air through the tube. Both the pump and the tube can be attached to the clothing of an applicator or worker, or be positioned in a stationary location. After the workday, the tube can be checked for the time weighted average of exposure. Exposure levels above an established threshold limit is an indication to take measures (i.e., shorten work shifts, increase ventilation in the area) to maintain the exposure level below the threshold level necessary to protect workers.

When using gas detector tubes, be aware that:
- Tubes will deteriorate with age. Some tubes have

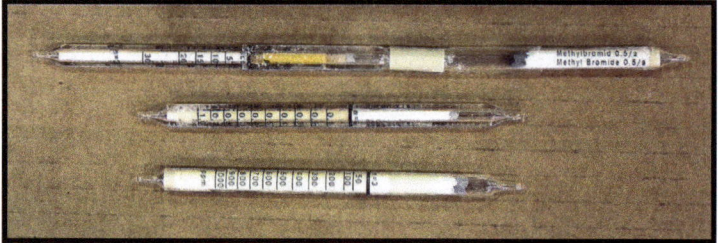

Figure 6: Monitoring with detector tubes. A worker using a detector tube to check for phosphine gas leaks from a grain bin (Top). Glass detector tubes are "fumigant specific" meaning you need specific tubes to detect each type of fumigant you use (Bottom).(Photo credits: (top) © Betsy Danielson, Iowa State University Extension and Outreach; (bottom) © Garo Goodrow, Penn State).

a shelf life of two years when stored at room temperature. Deterioration is more rapid above 86°F. You should always check the expiration dates of tubes prior to using them.

- Direct sunlight can affect the properties of the tubes. Follow the instruction manual on the proper storage of the tubes.
- At low temperatures, around freezing or below, tubes may not give reliable readings. You should warm the tubes to room temperature prior to using them for best performance.
- Tubes may have cross-sensitivity to gases other than those for which they are designed. You should obtain this information from the manufacturer.

Monitoring Precautions

We already covered the precautions related to air monitoring for fumigants. Although these precautions are "common sense", it is best to review these precautions to ensure that fumigations are safe for applicators and the general public.

- Always follow label and applicator's manual directions for the types of detectors to use.
- Use the appropriate detector (low or high concentration detector) for the task.
- Follow the device manufacturer's instructions for proper use and calibration.
- Make sure the detection device is calibrated and working properly.
- When using a detector to monitor for leaks outside the enclosed structure, make sure you are wearing appropriate PPE (i.e., respirator) or have it nearby if needed. If you do not know the concentration in an area, you must wear PPE.
- Follow the same precaution in the bullet above when monitoring to see if the fumigant has been properly and completely aerated from inside an enclosed structure.

Notification Requirements

In California, many non-soil fumigants are restricted materials and require a restricted materials permit from the local County Agricultural Commissioner's (CAC) office. After obtaining a permit, you may be required to submit a Notice of Intent before you conduct the fumigation. For more information, review the Laws and Regulations Study Guide, 3rd Edition, Chapter 2 (California, 2020), or contact your local CAC's office.

Be aware of any or all applicable notification requirements before performing a fumigation. Notification requirements may also be required by the product label (Figure 7). If required, agencies such as the fire department, local health agency, and the police department may need to be notified prior to the fumigation. Provide these agencies with the following information:
- Names and telephone numbers of all appropriate personnel in charge.
- Location, date, and time of application.
- Product and chemical name for the fumigant(s) used.
- SDS and a copy of the label.
- Required PPE and safety equipment.
- Fire hazard rating of the fumigant.
- Fumigation Management Plan that was developed.

SECTION 14

NOTIFICATION REQUIREMENTS

14.1 AUTHORITIES AND ON-SITE WORKERS
As required by local regulations, notify the appropriate local officials (fire department, police department, etc.) of the impending fumigation. Provide the officials an MSDS and complete label for the product and any other technical information deemed useful. Offer to review this information with the local official(s).

14.2 INCIDENTS INVOLVING THESE PRODUCTS
Registrants must be informed of any incident involving the use of this product. Please call 1-800-438-6071 so United Phosphorus, Inc. can report the incident to Federal and State Authorities.

14.3 THEFT OF PRODUCTS
Immediately report to the local police department thefts of metal phosphide fumigants.

Figure 7: Notification requirements on a label. This section, taken from a phosphine fumigant product applicator's manual, shows that notification of local authorities may be required, especially in the case of accidents or theft of product. Note that the text refers to an "MSDS," which is now known as SDS.

Even if not required by state or local laws, notification is a "best management" practice and will help with responding to any accidents or issues that may result from the fumigation.

Chapter 3 Review Questions

Correct answers are given on page 175.

1. What fumigant uses the "CT concept" to calculate the dosage?
    a. carbon dioxide
    b. sulfuryl fluoride
    c. phosphine

2. Flexible tubes filled with water or sand that are used to seal tarpaulins (tarps) to the ground are called _____.
    a. runners
    b. snakes
    c. hoses

3. The gas detection device that measures a gas's absorption of infrared radiation is _____.
    a. a thermal conductivity analyzer
    b. an ambient air analyzer
    c. a halide leak detector

4. Which of the following is responsible for issuing restricted material permits for the use of a non-soil fumigant that is a restricted material?
    a. the California Environmental Protection Agency
    b. Department of Pesticide Regulation
    c. the local County Agricultural Commissioner

5. Which of the following will determine the volume of a simple structure?
    a. length x height x diameter
    b. length x width x height
    c. radius x height x width

CHAPTER 4
Fumigation Safety

LEARNING OBJECTIVES

- ☑ Explain how applicators and others (handlers, bystanders) may be exposed to fumigants.
- ☑ Describe common mistakes that lead to direct exposure to fumigants.
- ☑ Describe the concerns, signs, and symptoms of human exposure to fumigants.
- ☑ List measures you can take to minimize exposure to fumigants.
- ☑ Outline first aid measures to take if someone is exposed to a fumigant.
- ☑ Describe when and where to take air monitoring samples.
- ☑ Explain the fumigant air concentrations that require handlers and applicators to wear respirators.
- ☑ Describe actions to take when an applicator experiences sensory irritation from a fumigant.
- ☑ Describe situations when handlers and workers must leave the work area entirely.
- ☑ Specify who should and should not be present in areas being fumigated.
- ☑ Specify who should and should not be present during aeration of a site.
- ☑ Define the two types of buffer zones and how to determine their size.
- ☑ Define the types of buffer zones and how to determine their size.
- ☑ Specify who is permitted to be in a buffer zone.
- ☑ Explain the requirements on posting warning signs.
- ☑ Explain the importance of posting treated sites, who must comply with posting, and who is responsible for posting.
- ☑ Explain a Fumigation Management Plan (FMP), its importance, and when it is required.
- ☑ List and describe the elements of an FMP and resources that can help you prepare it.
- ☑ Specify who is responsible for verifying the accuracy of the FMP.
- ☑ Explain how long you must keep an FMP on file, where to keep the plan and who must have access to it.
- ☑ Describe the purpose of a Post-Application Summary (PAS) and describe its elements.
- ☑ Specify who must prepare the PAS and when it must be completed.

Safe Use of Fumigants

Treat fumigants with respect to protect yourself, your coworkers, and the general public from accidental exposure. Safety must be a top priority for you when using fumigants. The most important thing you can do to ensure personal and public safety is to read and follow the fumigant label instructions.

Terms to Know

The following are important terms to know from this chapter. They will be *italicized* in their first use. They are explained in the text and defined in the glossary at the end of this manual.

Buffer Zone	Fumigation Management Plan	Toxic
Certified Applicator-in-Charge	Post-Application Summary	Toxicity
Exposure	Risk	Trigger levels
		Warning Sign / Placard

Fumigation Safety

As we have been noting throughout the study manual, fumigants are some of the most toxic pesticides we use, and they need to be used with skill and care. Applicators learn and get trained in how to do this, however the general public does not. The sites you treat (grain bins, utility poles, sewer lines, etc.) are often located near where people live and work. Livestock and other animals (e.g., pets, wildlife) may also be nearby. In nearly all cases, people and animals may come into contact with items that have been fumigated or be in, or nearby, areas where the fumigant is being used. Your ability to apply fumigants safely is critical. You must protect the public and the environment from exposure to the pesticide. There are several important ways to protect others from fumigant exposure. Some of these practices have already been discussed in previous chapters and include:

- Reading, understanding, and following the label directions.
- Preparing and planning the fumigation well before application.
- Sealing the fumigation site.
- Securing the fumigation site.
- Posting *warning signs.*
- Monitoring the fumigant.
- Safely transporting, storing, and disposing of fumigants and their containers.
- Properly aerating the treatment area.

Toxicity, Risk, and Exposure

Before diving into the specific human health concerns of fumigants we need to review some terms and concepts regarding toxicity of pesticides in general. Some pesticides are highly *toxic* to people—a few drops in your mouth or on your skin can cause extremely harmful—or even lethal—effects. Other pesticides are less toxic but with prolonged or substantial exposure, can also cause harm. Let us define a few terms to better understand the toxicity of the pesticide you will be using, the ways you may be exposed to the pesticide, and the risks of being exposed to the pesticide.

- *Toxicity* is a measure of the capacity of a pesticide (or other substance) to cause injury. That injury can occur soon after the exposure (acute) or long afterwards (chronic). Toxicity is a characteristic of the pesticide itself—some pesticides are very toxic while others are less toxic.
- *Exposure* (to a pesticide) occurs when you come into contact with a pesticide either through your skin, in your eye, ingesting it, or inhaling it (or its vapors). Unlike toxicity, you can have control over your exposure to the pesticide.
- *Risk* is a measure of the likelihood that a person will be harmed by the pesticide and its particular use. It is a product of both the pesticide's toxicity and the amount of exposure. Risk can be expressed by this formula:

RISK = TOXICITY x EXPOSURE

The risk (or chances of) of being harmed by a pesticide depends both on the toxicity of a pesticide and the level of exposure to that pesticide. If, for example, a pesticide is very toxic, but you have no exposure, you have no risk of being harmed. Keep in mind that any pesticide carries some toxicity, which means there is always a risk.

How People are Exposed to Fumigants

Fumigant applicators, other workers, and the public can potentially be exposed to fumigants. It is your job as an applicator to take measures to minimize exposure to the fumigant (including for yourself, other workers, and the public) and thus, minimize the risk of harm from the pesticide. While many pesticides can cause damage upon contact with the skin and eyes, the most harmful toxic effects usually occur when pesticides enter the body. The four main routes of exposure to pesticides are:

- Dermal exposure (when you get a pesticide on your skin),
- Oral exposure (when you swallow a pesticide),
- Ocular exposure (in your eyes), and

- Inhalation or respiratory exposure (when you breathe in pesticide vapors or dusts).

Risk to Handlers

With fumigants, respiratory exposure is the most common and hazardous route of exposure. The risk of inhaling a fumigant pesticide is greatest:
- When opening and releasing the fumigant,
- During application,
- During aeration of fumigation enclosure, and
- When disposing of spent solid fumigants (e.g., aluminum and magnesium phosphide).

> **Physical Condition and Fumigation**
>
> Make sure that you are in good physical condition if you actively take part in fumigation procedures.
>
> - Have a physical examination at least once a year, or more often if health conditions require them.
> - Do NOT participate in a fumigation if you suffer from colds or other respiratory problems that make breathing difficult.
> - Do NOT participate in a fumigation while you are undergoing medical treatments unless authorized to do so by medical personnel.

Always Work in Pairs

One of the most important ways a handler can protect themselves from fumigant exposure is to always work with another person. Another person can assist the handler if he or she is injured or exposed to fumigant vapors. Fumigant labels often require the presence of two people trained in applying the fumigant when the potential for exposure is greatest, such as during fumigant introduction, reentry into the fumigated structure before aeration, initiation of aeration, and during reentry when testing for clearance.

Risk to Handlers and Bystanders

Bystanders, as well as handlers, can also be at risk of exposure to fumigants. Many situations where the applicator did not read or follow the directions and safety precautions specified on the label have resulted in mistakes that led to unnecessary and harmful exposures. Some of these mistakes include:
- Neglecting to post fumigant warning signs at the treatment and/or buffer areas to warn people of possible hazards.
- Not using the appropriate PPE.
- Using poorly fitted respirators and/or failing to change respirator cartridges when required.
- Failing to properly seal the fumigation site.
- Neglecting to seal off connected rooms or structures that should not be fumigated.
- Failing to adequately aerate the commodity or space.
- Failing to monitor fumigant levels or conduct air monitoring (or not using the proper detectors) during aeration.

This list could likely go on for pages, but the basic principles are to always put safety first, be knowledgeable about the fumigants and equipment being used, be aware of all potential factors that could lead to exposure, and always follow label and applicator's manual directions.

Signs and Symptoms of Fumigant Exposure in People

As with any pesticide, exposure to fumigants can cause both acute and chronic effects in people. Acute effects of pesticide exposure are easier to detect, observe, or study than chronic or delayed effects. With acute effects, it may be easier to correlate symptoms to an exposure. Trying to determine if symptoms that occur years after an exposure are due to a specific pesticide exposure is much harder.

General Symptoms

Fumigant inhalation symptoms may be mild, moderate, or severe.
- Mild inhalation exposure can cause a feeling of sickness, ringing in the ears, fatigue, nausea, and tightness in the chest. Moving the person to where there is fresh air may help relieve these symptoms.
- Moderate inhalation exposure can cause weakness, vomiting, chest pain, diarrhea, difficulty breathing, and pain just above the stomach.
- Severe inhalation exposure symptoms can occur within a few hours to several days after exposure. Severe poisoning may result in pulmonary edema (fluid buildup in the lungs). This can lead to dizziness, blue or purple skin color, unconsciousness, and even death.

Symptoms for Specific Fumigants

Besides the general symptoms above, each fumigant active ingredient may have its own specific symptoms.

Methyl Bromide

Methyl bromide is particularly dangerous because it is both highly toxic and odorless. Because it is odorless, you will not be able to tell if respirator cartridges are spent unless the fumigant is used with a warning agent. It is nonirritating to the skin and eyes during exposure but later can cause serious skin or eye injury. You may not be able to detect toxic levels of the fumigant and can inhale dangerous levels of the fumigant without warning.

Contact with methyl bromide can cause severe chemical burns of the skin, respiratory tract, and other exposed tissue; delayed chemical pneumonia; and severe kidney damage. Any of these effects can be fatal. Victims also may experience extreme nervousness.

When small amounts of methyl bromide are inhaled, symptoms may be similar to alcohol intoxication. Methyl bromide exposure effects are also cumulative, where repeated low dose exposures can result in the buildup of the chemical in body tissue. These long term exposures may cause skin rashes, sometimes followed by mental confusion, double vision, tremors, slurred speech, and a lack of coordination.

Chloropicrin

As noted in Chapter 1, chloropicrin is mainly used in non-soil fumigations as a warning agent in structural fumigation, the Structural Pest Control Board in California issues the license for this type of use. The other non-soil use is to fumigate in-service utility poles, a use not currently registered in California. These non-soil fumigant uses of chloropicrin are outside the scope of this study manual.

Phosphine

Symptoms of phosphine exposure can be severe, but they are clearly recognizable and may be reversible if exposure is ended as soon as symptoms begin to appear. The primary route of exposure to phosphine gas is through inhalation. Phosphine is not absorbed through the skin.

> **Acute and Chronic Poisoning Terminology**
>
> Poisoning is poisoning, right? Not necessarily. The following are some terms that sound similar—and are related—but do have different meanings:
>
> - *Acute toxicity* is a measure of the capacity of a pesticide to cause injury as a result of a single one-time exposure.
> - *Acute exposure* is exposure to a single one-time dose of a pesticide.
> - *Acute effects* are symptoms that usually occur within minutes or hours after exposure to a toxic substance. These effects can be dermal, oral, or inhalation toxicity.
> - *Chronic toxicity* is a measure of the capacity of a pesticide to cause injury as a result of small, repeated exposures over a long period of time.
> - *Chronic exposure* is repeated exposure to a pesticide over a long time (usually years).
> - *Chronic effects* are those that appear over time and often a long time after exposure (months to decades). They are caused by repeated exposures to small pesticide exposures that do not immediately cause acute effects.

Slight or mild symptoms of phosphine gas exposure include fatigue, buzzing in the ears, nausea, pressure in the chest, and uneasiness.

Medium to heavy poisoning symptoms include general fatigue, nausea, gastrointestinal symptoms (vomiting, stomach ache, diarrhea), disturbance of equilibrium, strong pains in the chest, back pains, a feeling of coldness, and dyspnea (difficult or labored breathing).

Unlike methyl bromide, phosphine does not build up in body tissues. Any phosphine entering the body will be eliminated within 48 hours.

Severe poisoning rapidly results in severe dyspnea (difficulty breathing), cyanosis (blue or purple skin color resulting from inadequate oxygen), agitation, unconsciousness, and death. Death may be immediate, or it can follow several days of chemical pneumonia, paralysis of the central respiratory system, and/or edema (fluid buildup) of the brain. A victim's breath and vomit will have a garlic like odor.

Sulfuryl Fluoride
Symptoms of sulfuryl fluoride poisoning include depression, slowed gait, slurred speech, nausea, vomiting, stomach pain, exhibiting signs of inebriation, itching, numbness, twitching, and seizures. Inhalation of high concentrations can cause respiratory tract irritation or respiratory failure. Skin contact with the gas may not pose a significant hazard but contact with liquid sulfuryl fluoride can cause pain and frostbite due to rapid vaporization. Repeated or prolonged exposure to high concentrations can cause injury to the lungs and kidneys, weakness, weight loss, anemia, bone brittleness, stiff joints, and general ill health. Sulfuryl fluoride does not accumulate in body tissues.

MITC and MITC-Generating Fumigants (Metam Sodium and Dazomet)
The risk of inhalation exposure to metam sodium and dazomet themselves is slight. However, since metam sodium and dazomet decompose to MITC, carbon disulfide (CS_2), hydrogen sulfide (H_2S), and other gaseous products, these fumigants must be used with extreme caution. MITC is caustic and irritating to the eyes, respiratory mucous membranes, and the lungs. Inhalation of MITC can cause pulmonary edema that may result in severe respiratory distress, and coughing up of bloody, frothy sputum. Because our skin and mucous membranes are moist, metam sodium will produce MITC upon contact with skin and can cause skin and eye irritation.

Sulfur Dioxide (SO_2)
SO_2 exposure can cause nose, eyes, throat, and lung irritation. Typical exposure symptoms include sore throat, runny nose, burning eyes, and cough. Inhaling high levels of the fumigant can cause swollen lungs and difficulty breathing. Contact with SO_2 vapor can cause skin irritation or burns. Liquid SO_2 is very cold and can severely injure the eyes or cause frostbite upon contact with skin.

Carbon Dioxide (CO_2)
High concentrations or prolonged exposure to CO_2 can produce numerous health effects. These include headaches, dizziness, restlessness, a tingling or "pins and needles" feeling, difficulty breathing, sweating, tiredness, increased heart rate, elevated blood pressure, coma, asphyxia, and convulsions.

Carbon dioxide gas is an asphyxiant, a substance that decreases or displaces the normal concentration of oxygen we need in the air we breathe. Concentrations of 10% or more can produce unconsciousness or death in 1 minute or less.

Steps to Minimize Fumigant Health Effects
Practicing common sense goes a long way to protect yourself and the public from the risk of exposure to fumigants. Many of these common sense practices were already discussed earlier in this chapter ("How People are Exposed to Fumigants"). Make sure that you read, understand, and follow the product label and applicator manual and use the fumigant properly. Seal the site correctly. Use properly calibrated devices as indicated. Wear proper PPE which often includes a respirator. PPE, including respirators, will be discussed in detail in Chapter 5.

First Aid for Fumigant Poisoning
Providing first aid, while vital in helping a victim, is merely the initial response to a pesticide exposure. First Aid procedures are not, and should not be, a substitute for professional medical help. Be able to recognize when someone needs medical attention, know where you can get it and be prepared to get it. Seek medical attention when someone:
- Exhibits any illness or injury while, or soon after, working with pesticides,
- Exhibits any illness while in, or soon after being in, an area treated with pesticides,
- Exhibits symptoms of poisoning or injury exposure to a pesticide by any route,
- Ingests a pesticide, or
- Gets a pesticide in their eyes.

There are situations where medical attention is required even if symptoms are not observed. For example, a person who enters a fumigated space without the required respiratory protection should be immediately taken to a medical facility for evaluation and monitoring. The onset of symptoms from fumigant exposure can be delayed. Acute symptoms for sulfuryl fluoride may not be evident until 24 hours post-exposure while it may take up to 48 hours for symptoms to show with methyl bromide.

You should immediately seek medical help when faced with any of the situations described above. However, you may need to provide first aid until medical care can be given. Be sure to first protect yourself (i.e., wear required PPE) before trying to remove someone from an area under active fumigation or aeration. Get yourself and the other person to fresh air as soon as possible. The First Aid section of the pesticide label will give you first aid instructions specific to the fumigant involved.

California law requires an employer to make certain that any employee, who has a pesticide illness or has been exposed to a pesticide that might lead to an illness or injury, is immediately taken for medical care. You should never send an injured or ill employee alone to seek medical care. Always have someone who can provide medical personnel with information relating to the exposure accompany the victim. When transporting a victim of fumigant exposure, roll down the vehicle's windows to prevent the vehicle occupants from being exposed to fumes that may be released from the victim's body or clothing. Be ready to provide medical personnel with the U.S. EPA Registration number, full trade name, active ingredient and SDS of the fumigant used. Providing the product label may be helpful but be sure that the label is not contaminated, as this can be hazardous to the medical staff.

Figure 1. The national pesticide control center number. Call this number to be connected to the nearest poison control center in your area.

Where to Get Help

In California, your employer must post in a prominent place at the worksite, or work vehicle if there is no designated worksite, the name, address, and telephone number of a facility able to provide medical care. If the label requires a Fumigation Management Plan (FMP), the FMP must also include the contact information of local medical facilities. Also, the First Aid section of the pesticide label includes the telephone numbers of the manufacturer and Poison Control Center. The Poison Control Center is available 24 hours and is a good source of information on the treatment of pesticide poisonings. Calling 1-800-222-1222 will automatically direct a caller to the closest Poison Control Center in the area (Figure 1). While other hospitals and medical facilities may have some information, Poison Control Centers have the most complete and current information on responding to poisonings, and have specialized personnel trained to address and assist medical personnel with poisoning cases.

Air Monitoring

We covered air monitoring equipment in the previous chapter and distinguished between monitoring fumigant levels for efficacy vs. monitoring for safety. In this chapter, we will focus on monitoring fumigant levels for safety, specifically to protect fumigant handlers and bystanders. We will cover when air monitoring is required, where and when to take samples, when you might need to wear a respirator, and other issues related to air monitoring.

When to Monitor

The *certified applicator-in-charge* must conduct air monitoring as required by the product label. Air monitoring requirements will vary by product and type of fumigation. Read the label and applicator's manual for detailed air monitoring requirements. Below are some examples of when air monitoring is required:

- For methyl bromide, air monitoring samples need to be collected anytime someone wears an air-purifying respirator. Samples are collected at least every hour to ensure that the 5.0 ppm upper protection limit of the air-purifying respirator and its respirator cartridge are not exceeded.
- For sulfuryl fluoride, applicators must monitor the perimeter of the fumigation area, especially downwind, during release and aeration to ensure that the concentrations outside the fumigation area are kept within acceptable levels.
- For aluminum and magnesium phosphide products, monitoring must be conducted to determine the exposure of applicators releasing pellets into the treatment area. Applicators also need to monitor airborne phosphine concentrations in all indoor areas accessed by them, and other workers, during fumigation and aeration. Spot checks must be performed, especially if conditions change significantly, if a garlic odor is detected, or if a change in phosphine level is suspected. Some phosphine labels may also require monitoring the perimeter of

the fumigation area, especially downwind, to ensure that phosphine concentrations outside the fumigation area are within acceptable levels.

The examples above are not an exhaustive list of when air monitoring is required. Always read the label and applicator's manual of the product you are using for detailed information on air monitoring.

Where to Monitor

The label and applicator's manual also dictates where you need to take an air sample. Many labels indicate that air samples need to be taken "in the breathing zone" of applicators or handlers. The Operator Breathing Zone (OPZ) or Worker Breathing Zone (WBZ) is a hemispheric area in front of the face from which air samples are usually drawn, to get the best representation of what the worker is inhaling.

Besides breathing zones, the label and applicator manual may require air monitoring in the following locations:
- The area surrounding the vacuum chamber.
- Around the perimeter of the fumigation area (especially downwind) during release and aeration. This may involve walking around the fumigation area or structure with a personal monitoring device to determine whether excessive amounts of fumigant are escaping from a specific area.
- Areas around the tarped stack when fumigation workers without proper respiratory protection (SCBA) are present within the structure.
- Around the grain bin to confirm any leakage is not above allowable limits in an area that would affect nearby workers or bystanders.
- When using aluminum or magnesium phosphide, in all indoor areas where fumigators and other workers have access during fumigation and aeration.

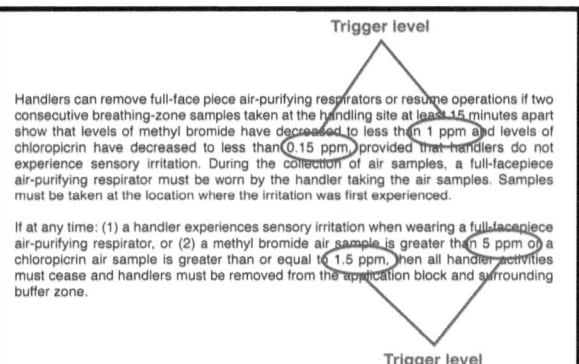

Figure 2: Trigger levels on a label. An example of a trigger level found in the "Respiratory Protection and Stop Work Triggers" section of a phosphine label. In this example, the required action when the Short Term Exposure Limit (STEL) for phosphine is unknown or exceeds 1 ppm for 15 minutes (the trigger) in a site under fumigation is to wear a Self-Contained Breathing Apparatus (SCBA).

When Respirators Are Required

Fumigant labels will provide direction for respirator use. The label also lists the exposure limit or the maximum fumigant concentration in which you can work without respiratory protection. This topic will be discussed in detail in Chapter 5, "Personal Protective Equipment."

Sensory Irritation

Sensory irritation may occur when people are exposed to fumigant vapors. Sensory irritation is a physical reaction that occurs when a fumigant reaches a certain concentration in the air. Sensory irritation can include burning, or irritation, of the eyes, nose, or mucous membranes. The label specifies the action that a fumigant handler (or other worker) must take if they experience sensory irritation, including one of the following:
- Use of an Air-Purifying Respirator (APR) to complete the task (if not already wearing one),
- Change to a full-face APR if experiencing sensory irritation while using a half-face APR, or
- Stop work activities and leave the application area.

Determination of Trigger Levels

Trigger levels are based on EPA exposure data and are used to prevent fumigant handlers from being exposed to the Maximum-Use Concentration (MUC). The MUC is the greatest air concentration of a hazardous substance (such as a fumigant) from which a person can expect to be protected while wearing a respirator.

The MUC is determined by the assigned protection factor of the respirator or class of respirators and the exposure limit of the hazardous substance.

The term "trigger levels" is found on some fumigant labels. The term "MUC" is not.

Trigger Levels

A *trigger level* is a specific air concentration of fumigant that prompts or "triggers" a required action (Figure 2). Fumigant labels may specify trigger levels that require applicators to either use a respirator or leave the work area.

When Workers Must Leave the Application Area
All applicators and workers must stop work and leave the application area when:
- The certified applicator-in-charge makes the decision to not have handlers wear an APR after experiencing sensory irritation, or
- A fumigant handler experiences sensory irritation while wearing a respirator, or
- Air monitoring samples shows fumigant concentrations above the Maximum-Use Concentration (MUC) for the respirators being worn.

Secure the Fumigation Site
Only trained and authorized pesticide handlers wearing the appropriate PPE are allowed into the site during the application and when aerating. One way to ensure this is to secure the site. You should lock all entrances to structures during fumigation. Some fumigant labels require you to use secondary locks to further guard against unauthorized entry (Figure 3). Employing a security guard to secure the area may be prudent under extremely sensitive conditions, such as fumigating a building in the middle of a densely populated area such as a town. Always inform other people who regularly use the building about the fumigation.

Buffer Zones
For some fumigants, the product label or restricted material permit conditions may require you to establish *buffer zones* to increase protections for anyone not protected by appropriate PPE (e.g., unprotected workers and bystanders). For non-soil fumigations, there are both treatment buffer zones **and** aeration buffer zones. For either zone, the certified applicator-in-charge must prohibit entry to the buffer zone by any person not involved in the application.

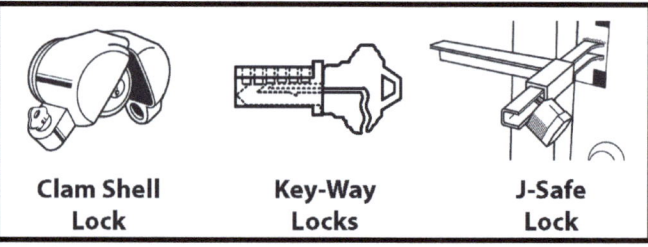

Figure 3: A secondary lock. You might need to use different kinds of locks to prevent entry to the fumigation site. Examples of locks you can use include clam shell, key way, and J-Safe locks. Clam shell locks are designed to prevent someone from using a key to unlock and enter a door.

Key-way locks work the same way as clam shell locks by inserting a two-part locking key into the door keyhole and removing only half of the key. (Top photo © Bonnie Rabe).

The treatment buffer zone extends from the perimeter of the treatment area to a specified distance. The treatment buffer zone prohibits entry from the time the fumigant is introduced into the fumigation enclosure and ends when aeration begins.

Similarly, the aeration buffer zone extends from where the fumigant is released in the treatment area (e.g., exhaust stack or building edge) to a specified distance. The aeration buffer zone begins at the start of aeration and ends when the air concentration of the fumigant in the area is below a label specified concentration.

Methyl bromide fumigant labels require the certified applicator-in-charge to establish and maintain buffer zones around the fumigation site. The U.S. EPA has a website that provides information on the buffer zones of six methyl bromide products. The applicable buffer zone from the website is part of the directions for use of these six products and must be included in the site-specific Fumigation Management Plan. In addition to label requirements, county permit conditions may have additional treatment or aeration buffer zone requirements. Buffer zones in restricted material permit conditions issued by the CAC are based on look-up tables found on recommended permit conditions supplied by DPR. The most restrictive requirement takes precedence and should be followed in instances where the label and CAC issued permit

> **IMPORTANT!**
> **A Note on Buffer Zones**
>
> At time of this writing, the only non-soil fumigants that are required to have buffer zones are methyl bromide products. Most soil fumigants are already required to have buffer zones.
>
> However, regulations do change so make sure you know whether you need to determine and monitor buffer zones for the product you are using. The label will tell you.

conditions are in conflict.

Buffer zone requirements can change to address hazards to handlers, nearby workers, other bystanders, and the environment. Always read carefully the product label and permit conditions of a fumigant to determine requirements for establishing and monitoring buffer zones.

Determining Buffer Zone Size

Both the treatment and aeration buffer zones are specific to each fumigation. The buffer zone size depends on the application rate, method of aeration, air-exchange rate, and other factors. If a fumigant product requires a buffer zone, refer to the label or permit conditions for instructions on how to determine the size of the buffer zones for each fumigation. For example, methyl bromide labels direct applicators to the U.S. EPA website to determine the required buffer zones for each type of fumigation. The minimum treatment or aeration buffer zone for methyl bromide is 10 feet.

Posting Warning Signs

Before fumigating, you must clearly post warning signs to discourage entry into the application site. *Warning signs* (also called *warning placards*) inform the public during fumigation and aeration that the site is being treated. Placards must be placed on the exterior sides of the treatment area and at all entrances.

Fumigant labels, California regulations, or restricted material permit conditions may have specific requirements for warning signs. These requirements include, but are not limited to, what the warning signs should look like, how the signs should be worded, when and where the signs should be placed, and under what conditions the signs can be removed (Figure 4). Follow these requirements explicitly. The certified applicator-in-charge is responsible for making sure warning signs are posted.

Reentry and Removal of Warning Signs

Workers without appropriate PPE can only enter a fumigated site after a certified applicator has determined that the area has been properly aerated or ventilated, and that the concentration of fumigant is at, or below, the safe level

---

16. PLACARDING OF FUMIGATED AREAS

All entrances to the fumigated area must be placarded including areas containing rodent burrows being fumigated (See Section 26.1). Placards must be made of substantial material that can be expected to withstand adverse weather conditions, and must bear the wording as follows:

1. The signal words DANGER/PELIGRO and the SKULL AND CROSSBONES symbol in red.
2. The statement, "Structure and/or commodity under fumigation. DO NOT ENTER/NO ENTRE".
3. The statement, "This sign may only be removed by a certified applicator or a person with documented training after the structure and/or commodity is completely aerated (contains 0.3 ppm or less of phosphine gas)."
If incompletely aerated commodity is transferred to a new storage structure, the new structure must also be placarded if it contains more than 0.3 ppm. Workers exposure during this transfer must not exceed allowable limits.
4. The date the fumigation begins.
5. Name and EPA registration number of fumigant used.
6. Name, address and telephone number of the Fumigation Company and/or applicator.
7. A 24-hour emergency response telephone number.

All entrances to a fumigated area must be placarded. Where possible, place placards in advance of the fumigation to keep unauthorized persons away. For railroad hopper cars, placards must be placed on both sides of the car near the ladders and next to the top hatches into which the fumigant is introduced. Do not remove placards until the treated commodity or area is aerated down to 0.3 ppm hydrogen phosphide or less. To determine whether aeration is complete, each fumigated structure or transport vehicle must be monitored and shown to contain 0.3 ppm or less phosphine gas in the air space around and, if feasible, in the mass of the commodity.

Figure 4: Warning sign or placard requirements on a label. The label will list what a placard should say, look like, where it should be posted, and more.

Warning Sign or Placard?

Sometimes terminology can be confusing. If a fumigant label tells you to post a warning sign, is the requirement the same as posting a placard?

Some people associate a "placard" to a diamond shaped sign with a symbol(s) indicating flammability, toxicity, or other dangerous hazards (right). This may be true in some circumstances. However, when transporting dangerous chemicals, vehicles have to carry placards like those shown on the right. Fumigant labels may use "placard" and "warning signs" interchangeably—sometimes even in the same sentence as in this example from a phosphine label:

"Placard all entrances to the treated spaces with fumigation warning signs."

So, what's an applicator to do? As always, follow the directions on the label. Even if the label uses both the terms warning sign and/or placard, it will specify the language to use. Regardless of whether the label refers to a posting as a sign or a placard, follow the guidelines on the label that detail requirements for the content and locations of these postings.

(Photo © Bonnie Rabe)

indicated on the label. It is also only then that the certified applicator (or certified applicator-in-charge, as indicated on the label) may remove or direct another worker to remove posted warning signs.

Fumigation Management Plans

Most fumigant products require you to develop a *Fumigation Management Plan* (FMP). The FMP is an organized, written description of the required steps involved to help ensure a safe, legal, and effective fumigation. It will also assist you and others in complying with pesticide product label requirements. An FMP helps the certified applicator-in-charge to compile necessary documents and information in advance, identify risks and hazards before beginning the treatment, and provide emergency protocols. Specific details of what needs to be included in an FMP will vary by product, and from one type of fumigant to another.

Common Headings Found in Fumigant Management Plans
- Planning and Preparation
- Personnel
- Monitoring
- Notification
- Site Prep and Sealing
- Application and Period of Fumigation
- Post-Application Operations

 REMEMBER! When the label requires an FMP, the FMP is considered a legal requirement. When you are required to prepare an FMP, you must develop it **before** you perform an application.

When developing an FMP, refer to the applicator's manual for specific requirements for the fumigant you are using. The certified applicator-in-charge is responsible for working with the property or site owner and/or the responsible employees of the site to be fumigated to develop an FMP and to ensure that the FMP is accurate, complete, current, readily available for review, and followed.

Some key requirements of an FMP include:
- Retaining the FMP and related documentation (i.e., monitoring records) on file for a minimum of 2 years.
- Having the FMP on-site and readily accessible during fumigation.
- Making the FMP readily available to emergency responders and to local, state, and federal enforcement personnel.

Elements of an FMP

The elements of an FMP can be found in the applicator's manual under "Required Written Fumigation Management Plan" or a similar heading. The required content of an FMP is typically organized into descriptive subheadings that often include a detailed list of tasks to consider under each of the subheadings. For example, the FMP must adequately characterize the site, and include appropriate monitoring and notification requirements. Because each fumigation is different, the FMP for each project will differ. If the same area is being fumigated a second time, you will need to update and modify the FMP for the new application. Below are some key elements of an FMP. As always, follow the label for specific requirements related to the FMP you are developing for a fumigation.

Planning and Preparation: In this section, you must determine and describe the purpose for the fumigation. As you fill out this section, it is an implicit acknowledgement that you have read and understood the pesticide label, applicator manual, and SDS documents which should be printed and available on site. Filling out this section also assumes that you have confirmed that the site, area, or structure is listed on the label, and have conducted an inspection to determine the suitability of the site, area, or structure for fumigation for pest control.

Personnel: This section identifies and lists all on-site personnel and includes written confirmation that all personnel in and around the structure to be fumigated have been notified **prior to application**. Make sure all fumigation personnel have read the applicator's manual and are aware of potential hazards and what to do in cases of an emergency.

Monitoring: This section ensures that proper monitoring equipment is in working condition and readily available. Many labels require that you keep a log of all monitoring activities and their outcomes.

Notification: In this section, you need to confirm that you notified the appropriate local authorities (e.g., CAC, fire or police departments) **prior to conducting the fumigation**. You will also need to prepare a written emergency response plan that describes the actions to take in case of an emergency, as well as list the names and telephone numbers of authorities and other people that need to be contacted, if needed.

Sealing Procedures: This section describes the proper sealing techniques you will conduct on the site/area to ensure that the fumigant remains within the fumigation site/area. The section also addresses the placement of warning signs at all entrances and sides of the fumigation site. If the site has been previously fumigated, you need to review the previous FMP to determine if any construction or remodeling has occurred since the last fumigation which may require modifying and updating your FMP.

Application Procedures and Fumigation Period: In this section, you attest to having the required number of trained and certified applicators on site (often at least two). This may also involve signifying that you will employ a security guard when entry into the fumigation site by unauthorized persons cannot otherwise be ensured (e.g., by secondary locks, barricades, etc.). When entering structures, always follow OSHA rules for confined spaces. If fumigating in-transit vehicles or containers, you will need to include documentation that the receiver has been notified of the fumigation. This is also the section where you describe the procedure to turn off, and address all possible sources of electricity, sparks, or flame, as required on the label.

Post-Application Operations: In this section, describe the procedures you will take to ensure that unauthorized people cannot get near the site/area during aeration. The section also addresses air monitoring procedures that you will conduct to determine fumigant levels outside an application site/area/structure, and determine fumigant concentrations in a fumigated structure, to clear or allow reentry. This section will also outline procedures to minimize bystander exposure post-fumigation such as turning on ventilating or aerating fans, the use of a sufficiently sensitive detection device to confirm that aeration is complete, and that the fumigated space has been cleared for reentry before the removal of fumigation warning signs. Always keep accurate records of both efficacy and safety air monitoring results to document the completion of aeration post fumigation.

Help with FMPs

FMPs can be complex. To assist you, many of the companies that manufacture fumigants provide templates for the FMP as do some government agencies. If you need more information or assistance for the FMP, you can contact the pesticide manufacturer or your local CAC's office.

Post-Application Summary

While the FMP describes the plan for conducting the fumigation, the *Post-Application Summary* (PAS) serves a different purpose. It describes:
- Any actions that occurred during the application that differed from the FMP.
- Measurements (e.g., humidity) taken to comply with Good Agricultural Practices (GAPs), if not recorded in the FMP.
- The National Weather Service forecast during application and 48 hours following application.
- Any incidents or complaints.

NOTE: At the time of this manual's writing, non-soil fumigant labels did not have information or requirements regarding the next section: "Post-Applicator Summary." In the future, U.S. EPA may call for similar requirements for non-soil fumigants. We present information below that presently applies to soil fumigations. As always, read the label to determine the requirements you need to follow.

PAS Contents

The following elements (if required) are included in a PAS:
- Application and application block details.
- Weather conditions.
- Tarp damage or repair.
- Tarp perforation or removal details.
- Information on complaints, incidents, equipment failure, or other emergency, and emergency procedures followed.

- Air monitoring results.
- Fumigant treated area and buffer zone posting and removal dates.
- Deviations from the FMP.

The certified applicator-in-charge must complete the PAS **within 30 days following the application**. As with the FMP, the certified applicator and the owner must keep a copy of the PAS for a minimum of 2 years.

Safety! Safety! Safety!

Any pesticide can be potentially harmful to people. Fumigants, by their very nature, can cause severe injury and death. As pest control professionals, we would like to think that we are careful and mindful during fumigations, but accidents can, and do, happen. Always read the label, follow label directions, and pay strict attention to all safety requirements. By doing this, you will protect yourself, your coworkers, and bystanders from harm.

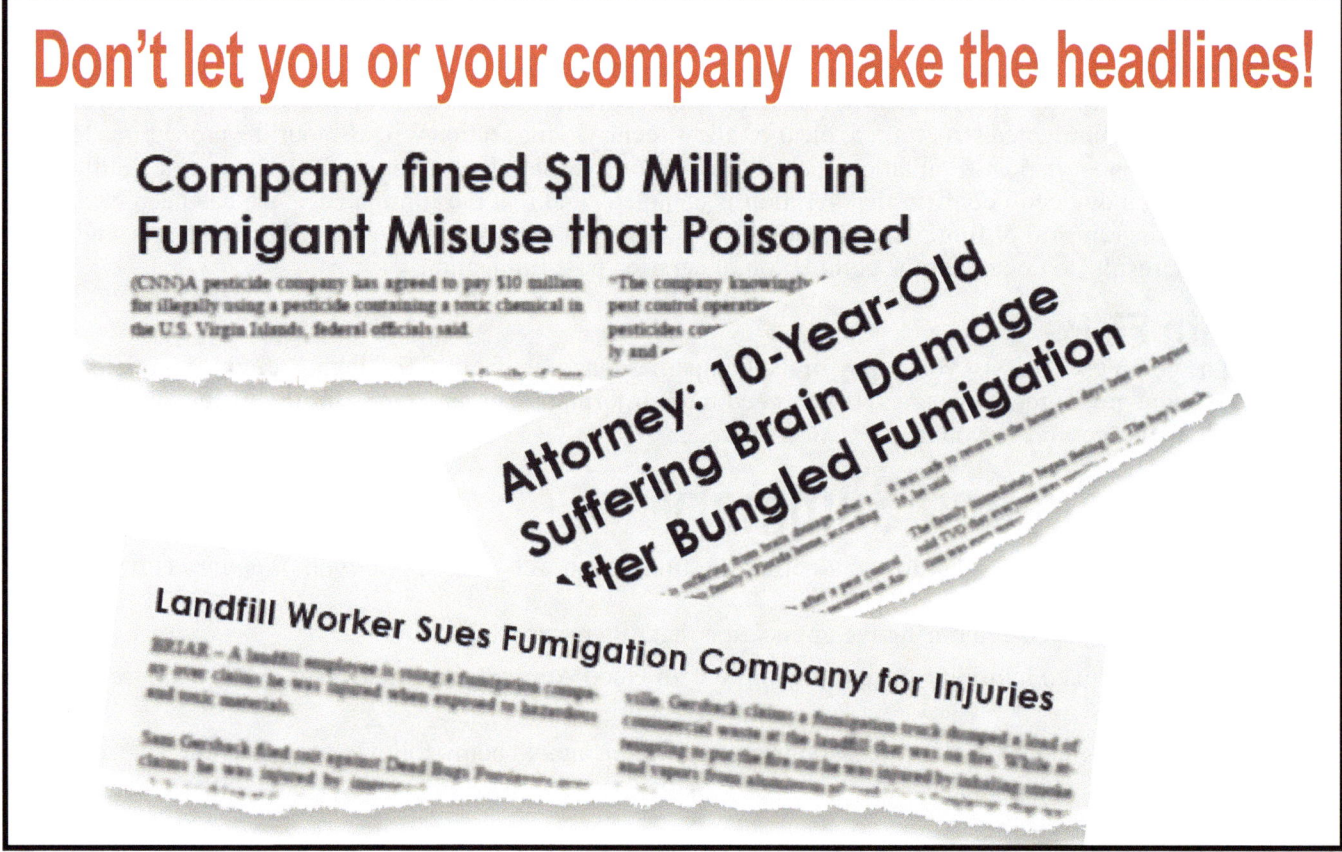

Table 4-1: Fumigant Practices in Non-Soil Fumigation

With a few exceptions, the same fumigant practices apply to the different fumigant types or methods. Table 4-1 summarizes some of these fumigant practices. Note that some fumigant practices apply to all the types or methods discussed in this study manual while a few are applicable to just some.

Table 4-1 is not an exhaustive list. Always follow the labeling requirements and be aware of the state and local regulations related to fumigant use.

Table 4-1: Fumigant Practices in Non-Soil Fumigation

FUMIGANT PRACTICE	CHAMBER (Chapter 7.1)	COMMODITY (Chapter 7.2)	BURROW (Chapter 7.3)	OTHER STRUCTURES (Chapter 7.4)	SPOT (Chapter 7.5)	TRANSPORT VEHICLES, CONTAINERS, SHIPS (Chapter 7.6)	SEWER ROOT LINE CONTROL (Chapter 7.7)	WINE BARRELS AND CORK (Chapter 7.8)	REMEDIAL WOOD PROTECTION (Chapter 7.9)
Have the Non-Soil Fumigation category	✓	✓	✓	✓	✓	✓	✓	✓	✓
If applying fumigants for hire, obtain PCB licensing from DPR	✓	✓	✓	✓	✓	✓	✓	✓	✓
Verify the fumigant is registered for the intended use	✓	✓	✓	✓	✓	✓	✓	✓	✓
Evaluate the use site and the application's impact on surrounding properties	✓	✓	✓	✓	✓	✓	✓	✓	✓
Determine if you need a restricted material permit from the County Agricultural Commissioner's office	✓	✓	✓	✓	✓	✓	✓	✓	✓
Determine if the label or permit conditions require buffer zones for the fumigation	✓	✓		✓	✓	✓	✓		
Prepare a site-specific written Fumigation Management Plan (when required)	✓	✓	✓	✓	✓	✓		✓	
Train employees on pesticide safety, including use and handling techniques	✓	✓	✓	✓	✓	✓	✓	✓	✓
Tightly seal the chamber	✓	✓		✓	✓	✓			
Wear the appropriate PPE, including respiratory protection (when required)	✓	✓	✓	✓	✓	✓	✓	✓	✓
Secure the site and place warning signs	✓	✓	✓	✓	✓	✓	✓	✓	
Have gas-monitoring devices available	✓	✓	✓	✓	✓	✓		✓	
After adding the fumigant, tightly seal the barrel or bag								✓	

Chapter 4 Review Questions

Correct answers are given on page 175.

1. The most hazardous route of exposure when using a fumigant is _____.
    a. oral
    b. dermal
    c. respiratory

2. Which fumigant accumulates in body tissues?
    a. methyl bromide
    b. sulfuryl fluoride
    c. phosphine

3. If you need to provide first aid to a victim who has been exposed to a fumigant, you should first _____.
    a. protect yourself by putting on the required personal protective equipment
    b. refer to the first aid section of the label for emergency medical contact information
    c. get the victim to fresh air to avoid the possibility of more exposure

4. Who is allowed to remove posted warning signs after a fumigation?
    a. the certified applicator-in-charge
    b. the owner of site or property manager
    c. the fumigation inspector

5. Which of these describes the written description of the required steps involved to help ensure a safe, legal, and effective fumigation?
    a. Emergency Response Plan
    b. Fumigation Management Plan
    c. Site Notification Plan

CHAPTER 5

Personal Protective Equipment

LEARNING OBJECTIVES

- ☑ Explain the importance of following label directions for personal protective equipment (PPE).
- ☑ List the PPE needed for fumigation.
- ☑ Compare and contrast PPE for different kinds of non-soil fumigants.
- ☑ Explain how to properly use PPE.
- ☑ Describe the different types of respirators used in fumigations. Compare these respirators to other respirators that are not designed for use in fumigations.
- ☑ Explain the requirements for replacing respirator cartridges or canisters.
- ☑ Explain the importance of a medical evaluation before using respirators.
- ☑ Explain when the use of a self-contained breathing apparatus (SCBA) is indicated and how to use it.

Terms to Know
The following are important terms to know from this chapter. They are explained and *italicized* in the text and defined in the glossary at the end of this manual.

Atmosphere-Supplying Respirator	Fit Test	Respirator
	Fit Check	Supplied-Air Respirator
Canister	Personal Protective Equipment	

Personal Protective Equipment

The principles and precautions regarding the use of pesticides also apply to the use of fumigants. However, there are additional considerations when using fumigants because they, and their uses, are so different from other types of pesticides. In this chapter, we will discuss some of the special concerns regarding the use of fumigants and *personal protective equipment* (PPE).

Clothing and devices (such as *respirators*) that protect you from contact with pesticides are called personal protective equipment or PPE for short. The label for each product lists the minimum PPE required for using that fumigant. Label PPE requirements may vary with the fumigant and with the task you are performing. California also has regulations specifying the minimum PPE that employees must use while handling pesticides. Restricted material permit conditions may also specify the required PPE when handling fumigants. You are legally required to follow all PPE instructions and to wear at least the minimum required PPE. Fumigants can be highly dangerous and even deadly, so it is extremely important for you to follow label directions and California regulations to protect yourself from harm.

Differences Between PPE for Fumigant and Non-fumigant Use
Fumigants are distinct from other pesticides since they are toxic as gases. Because of this, the PPE requirements for handling fumigants are often quite different than those for handling non-fumigant pesticides. Read the pesticide label for the PPE requirements specific to the handling activity you will be performing. Some examples of the differences in PPE requirements between fumigants and non-fumigant pesticides include:
- Some labels may direct you to wear loosely woven work clothes and even short sleeves when handling fumigants. Tightly woven and tightly fitted clothes can trap fumigants next to your skin, which can lead to severe burns.
- Some labels specifically tell you NOT to wear items such as a chemical-resistant apron or spray suit.

- It is not always recommended that you wear chemical-resistant gloves or boots when handling fumigants. Methyl bromide labels tell you not to wear gloves at all. Furthermore, do not wear rubber boots because the rubber absorbs methyl bromide, which can then be trapped next to your skin and cause severe burns.
- Some fumigant labels may tell you to wear cotton gloves (which are normally prohibited for non-fumigant pesticides), while others will require the use of chemical-resistant gloves and footwear.
- Fumigant labels may be very specific about the type of eye protection to wear when a full-face respirator is not required. For example, methyl bromide labels require a full-face shield or safety glasses but prohibit the use of goggles.

General Precautions for Fumigants

Take the following precautions whenever you handle fumigants:
- PPE is effective only if it fits correctly and is used properly. The fumigant label and the manufacturer use instructions include directions for keeping PPE clean and properly maintained.
- Do not wear jewelry, watches, or other items that may trap fumigant gas against your skin.

Basic PPE

We will discuss respirators in more depth later in this chapter. Remember to always follow the label instructions regarding the PPE required for the product you are using. In California, employers are required to provide their employees with the required PPE. Your employer is also responsible and required to inspect and clean PPE daily, and repair and replace PPE when needed. Your employer must also make certain that PPE is stored in a pesticide free place. You must properly wear the required equipment provided by your employer. Never take PPE into your home.

Certain PPE must be chemical-resistant. Chemical-resistant PPE is made of a material that does not allow any measurable movement of the pesticide through it during use. For example, a pesticide label might direct you to "Wear butyl rubber or barrier laminate gloves," which are examples of chemical-resistant PPE.

Protective Clothing

Some fumigants packaged in cylinders require you to wear long-sleeved shirts and pants during fumigant introduction. Other products may direct you to wear short sleeves and loose-fitting clothing. In California, employers are required to make certain that employees wear coveralls (one- or two-piece garment that cover the entire body except the head, hands, and feet) when they handle non-fumigant pesticides that have the signal word "DANGER" or "WARNING." However, for a fumigant with the same signal words, employers are only required to make certain that employees wear coveralls **if** the label explicitly requires it. The exception to this would be the application of solid fumigants such as aluminum phosphide, magnesium phosphide, and smoke cartridges. Coveralls must be provided by the employer as PPE and should not be confused with work clothing that can be required to be provided by the employee.

Some fumigant labels may require you to wear a chemical-resistant apron or chemical-resistant footwear when handling fumigants (e.g., when performing tasks where there is potential of coming into contact with a liquid fumigant). California requires chemical-resistant aprons to cover the front of the employee's body from the mid-chest to the knees. Chemical-resistant footwear includes chemical-resistant shoes, chemical-resistant boots, or chemical-resistant coverings worn over shoes or boots unless the pesticide label specifies something different.

Eye Protection

Fumigant labels for products contained in pressurized cylinders often require eye protection, such as goggles or a full-face shield. Eye protection is worn during fumigant introduction to protect your eyes should the introduction hose accidentally burst or disconnect from the cylinder.

California requires employers to make certain employees wear protective eyewear in many situations when handling pesticides, even if the pesticide label does not. See DPR's Laws and Regulations Study Guide, 3rd Edition (California, 2020), or contact your local County Agricultural Commissioner's office for more information.

Gloves

The need for gloves varies with the product. For example, some fumigants require you to wear gloves because of possible skin irritation. Other fumigants, particularly liquid products, do not require gloves. Some may even prohibit you from wearing gloves. Some fumigants, such as methyl bromide, can cause serious injury to the skin if clothing or jewelry holds the gas tight against the skin. This is the reason that gloves are often not recommended for fumigant application. As always, consult the label to determine glove requirements and what precautions you should take.

In many situations the pesticide label may not require the use of chemical-resistant gloves, but California regulations require employers make certain employees handling pesticides wear them. If the product label does not specify a barrier material or category, the gloves can be made from barrier laminate, butyl rubber, nitrile rubber, neoprene, natural rubber, polyvinyl chloride, or Viton®. Refer to DPR's Laws and Regulations Study Guide, 3rd Edition (California, 2020), or contact your local CAC's office for more information.

PPE Exemptions
California regulations allow certain exceptions and substitutions to PPE required by the product label or listed in regulation. For example, protective eyewear is not required when applying solid fumigants (i.e., aluminum and magnesium phosphide, and smoke cartridges) to vertebrate burrows. However, the employer must make certain the exempted PPE is present and available at the worksite. Refer to DPR's Laws and Regulations Study Guide, 3rd Edition (California, 2020), or contact your local CAC's office for more information.

PPE According to Specific Task
PPE requirements for fumigants vary by the product and often, by the task you are performing. Fumigant labels may have one set of PPE requirements for fumigant handlers that perform tasks that will not expose them to contact with liquid fumigant, and a separate set of requirements for handlers whose tasks are more likely to result in contact with liquid fumigant.

For example, when using methyl bromide, fumigant handlers that are not expected to come in contact with a liquid fumigant must wear a long-sleeved shirt, long pants, shoes, and socks. When handlers are more likely to contact the chemical, they must also wear chemical-resistant gloves, a chemical-resistant apron, protective eyewear (not goggles), and footwear with socks.

Always read and follow the label directions of the fumigant you are using to protect yourself from harm, and make sure you wear the appropriate PPE corresponding to the activities you will be performing.

PPE for Specific Fumigants
We will now discuss the specific PPE requirements for each of the non-soil fumigants discussed in this study manual. Always read and explicitly follow the PPE requirements found on the most current version of the label, especially since regulations, and sometimes the label itself, may change. Remember, in addition to the wearing the required PPE to enter a treated space, you also need the correct air monitoring equipment to determine if the levels of the fumigant in the space requires PPE to safely enter the space. Air monitoring is just as important during aeration to determine if the space has been adequately aerated to not require the use of PPE.

PPE for Phosphine
Solid aluminum and magnesium phosphide react with moisture to release phosphine gas. Wear only light weight cotton gloves and loose-fitting clothing while handling or fumigating with these formulations to prevent residues from being trapped against the skin. If residues become trapped against the skin, they can react with moisture on the skin, releasing phosphine gas. Gloves must remain dry during pellet or tablet application. Gloves and clothing worn during fumigation should be aerated in a well-ventilated area prior to laundering. California regulations specify that leather gloves, specifically used to apply aluminum or magnesium phosphide are considered cleaned when they have been aerated for 12 hours or more.

For phosphine formulations in cylinders, the label may require leather work gloves or leather face cotton gloves. The label may also require safety glasses when working with pressurized equipment and recommend steel-toed safety shoes when handling cylinders.

Respiratory protection is required when phosphine levels are 0.3 ppm or higher. The use of a full-face respirator with a phosphine *canister*, or a *supplied-air respirator*, if phosphine levels are from 0.3 to 15 ppm. At levels above 15 ppm, or when levels are unknown, a supplied-air respirator is required. Keep in mind that there is no antidote for phosphine poisoning and exposure to even small amounts of phosphine can cause headache, dizziness, nausea, vomiting, diarrhea, drowsiness, cough, and more serious symptoms, even death.

PPE for Sulfuryl Fluoride

Wear splash resistant goggles or a full-face shield when handling liquid sulfuryl fluoride during introduction of the fumigant, or when working around any lines containing the fumigant under pressure. Do not wear gloves or rubber boots. Wear a loose-fitting or well-ventilated long-sleeve shirt, long pants, shoes, and socks.

A supplied-air respirator is required if the concentration of sulfuryl fluoride is unknown or over 1 ppm.

PPE for Methyl Bromide

Do not wear jewelry or gloves. Methyl bromide can cause serious dermal (skin) injury if clothing, jewelry, or gloves hold the gas tightly against your skin. When handling the fumigant or pressurized hoses, wear a splash resistant full-face safety shield.

A supplied-air respirator is currently required when the concentration of methyl bromide is above 5 ppm. Future label changes or regulations may require respiratory protection at a lower concentration or when prolonged exposure occurs.

PPE for Sulfur Dioxide

If using braided hoses, wear a NIOSH-approved full-face respirator with an acid gas or combination organic vapor/acid gas cartridge until the hoses are in place. In addition, wear a long-sleeved shirt and long pants, chemical-resistant gloves (butyl rubber or neoprene are examples), socks and chemical-resistant boots.

Some labels may require additional PPE in some situations. For example, when using application equipment that is not equipped with wire braided hoses, the label may require you to wear full-body protective clothing, gloves, and boots, all of which must be impervious to SO_2. In addition, you need to wear goggles or a face shield and approved respiratory protection.

Never wear contact lenses when working with SO_2 gas since the lenses will absorb and trap vapors. Vapors absorbed by a contact lens can cause serious eye irritation or injury. Should an exposure occur, contact lenses would also make decontamination of the eyes more difficult.

No respiratory protection is required if the concentration of SO_2 in the worker area does not exceed 2.0 ppm. If at any time the 2.0 ppm concentration is exceeded, all persons working in the fumigation area must wear a NIOSH-approved full-face respirator with an acid gas or combination organic vapor/acid gas cartridge. If the concentration is greater than 10 ppm or is unknown, all persons working in the area must wear a NIOSH-approved self-contained breathing apparatus (SCBA) or a supplied air respirator. No one should enter a high SO_2 concentration area using only a full-face respirator because these have a short-term and limited capacity for providing protection. It is not possible for someone wearing a full-face respirator in a high SO_2 concentration area to know if the absorbing capacity of the cartridges of the full-face respirator has been reached and if it is still providing adequate respiratory protection.

PPE for Carbon Dioxide

For handling activities in enclosed areas during and after application, use either a supplied-air respirator or a self-contained breathing apparatus (SCBA). After the application of CO_2 in enclosed spaces, aerate treated areas until the CO_2 level measured with commercially available monitoring equipment is below 5,000 ppm.

PPE for MITC and MITC-Generating Products

Handlers that use sewer line root control products containing metam sodium, and who are likely to directly contact the fumigant, may be required to wear splash resistant goggles or a full-face shield, chemical-resistant gauntlet type gloves and boots, and body covering including shirt and long pants or long-sleeved clothing. Mixers and loaders must also wear a chemical-resistant apron or cloth coveralls when a closed system is not used.

Handlers operating or monitoring application equipment must wear chemical-resistant footwear and body covering including shirt and long pants or long-sleeved clothing. Chemical-resistant gauntlet type gloves and, either a half-face respirator with organic vapor cartridges plus eye protection or a full-face respirator with organic vapor cartridges, must always be readily available and accessible to handlers. For utility pole applications, the required PPE may vary by product. Always read and follow the product label for guidance. For example:

- A metam sodium product labeled for this use requires handlers to wear coveralls over a long-sleeved shirt and long pants, socks, chemical-resistant shoes, goggles or face shield, and chemical-resistant gloves.
- A dazomet product labeled for this use requires handlers to wear long pants, a long-sleeved shirt, chemical-resistant gloves, and goggles or face shield.
- An MITC product labeled for this use requires handlers to wear a long-sleeved shirt and long pants, socks and chemical-resistant footwear, goggles or face shield, and chemical-resistant gloves.

Figure 1: NIOSH-Approved respirators. Be sure that you use a NIOSH-approved respirator when performing fumigation activities. Be aware that there are respirators on the market that are not NIOSH-approved, such as nuisance dust masks and some surgical masks.

Respirators

This section provides a brief overview of California's regulatory respiratory requirements to protect worker safety. Title 3 California Code of Regulations (3 CCR) section 6739 provides important details on these respiratory requirements. You can also contact your local CAC's office for more information on respirators. DPR also maintains a webpage on Respiratory Protection Compliance Assistance for employers.

For fumigants, respiratory (inhalation) exposure is the most common and hazardous route. A respirator is a safety device that covers at least a person's mouth and nose and protects the wearer from breathing in hazardous substances. Remember that fumigants are some of the most toxic pesticides and inhaling even small amounts of these pesticides can be fatal. Training on the safe, proper, and effective use of respirators is crucial. Respiratory protection is required when concentration levels of fumigants are unknown or above specified levels.

Employer Responsibilities for Respirators

Using respirators requires planning ahead. If the pesticide label, California regulation, county restricted material permit conditions, or employer workplace policy require the use of respirators, an employer **must** develop a formal respiratory protection program. This program must meet all the requirements in 3 CCR section 6739 including written worksite specific operating procedures for:
- Selecting respirators;
- Medical evaluations of employees;
- Fit testing procedures for tight-fitting respirators;
- Cleaning, storing, inspecting, repairing, maintaining, and replacing respirators;
- Procedures for air-supplying respirators (if applicable);
- Training employees on Immediately Dangerous to Life or Health (IDLH) atmospheres (if applicable);
- Training employees on the proper use of respirators; and
- Procedures for effectiveness evaluation.

After a written program is in place, the employer must:
- Work with a physician or other licensed health care professional to determine medical fitness of employees to wear a respirator, or any other conditions that the employee must follow to wear a respirator;
- Train employee(s) on the use of the respirator, and re-train them annually thereafter;
- Fit test employee(s) on the respirators they will be wearing prior to use, and annually thereafter;
- Maintain records of the program for three years;
- Document annual consultations with the employee(s) on the effectiveness of the program; and
- Annually review the program, making adjustments as necessary.

NOTE: DPR's Worker Health and Safety Branch has a guidance document for developing a California compliant respiratory protection program.

Respirator Use

Always use the type of respirator that is consistent with the label requirement. Never substitute a respirator with a type that does not comply with the label requirements. Carefully review the label for the respirator requirements to determine:

- Whether respiratory protection is required,
- The correct type of respirator for the fumigant in use, and
- Application specific situations when respiratory protection is required.

 All respirators used by fumigant applicators must be approved by the National Institute of Safety and Health (NIOSH (Figure 1)). The specific type of respirator required may vary depending on the health of the applicator, the type of fumigant used, and the conditions of its use.

Figure 2: Particulate respirators. Never use dust/mist respirators, including "N-95s", for any fumigant work. (Photo © UW-Madison Pesticide Applicator Training Program).

California PPE Exemptions for Solid Fumigants

In many situations, California regulations may require employers to make certain that their employees:
  a) Wear protective eyewear in many situations when handling pesticides, and/or
  b) Wear coveralls if the pesticide has the signal word "DANGER" or "WARNING".

However, an employer is exempt from these regulatory requirements if the employee will treat a vertebrate burrow with a solid fumigant, including, but not limited to, aluminum phosphide and magnesium phosphide, and smoke cartridges.

Types of Respirators

The two basic types of respirators are air-purifying respirators and atmosphere-supplying respirators. Table 5-1 (page 70) summarizes the characteristics of each respirator type.

Air-Purifying Respirators

Air purifying respirators pass ambient air through an air-purifying element (filters, cartridges, or canisters) to remove contaminants from the air before reaching the user. Contaminants include particles, gases, vapors, aerosols (droplets and solid particles), or a combination of these. Different types of air-purifying devices provide varying levels of respiratory protection.

Particulate respirators are the simplest, least expensive, but least protective of the respirators available (Figure 2). These respirators only filter out particles (dust, mists and fumes). They do not protect against gases or vapors and **will not** protect you from fumigants.

Chemical Cartridge/Canister Respirators (may also be referred to as elastomeric) include a facepiece or mask and a piece that filters out specific gases and vapors in the air through chemical filters called cartridges or canisters. These filters provide protection as long as their absorbing capacity has not been used-up. Straps secure the facepiece to your head. The headpiece can be a half-face, which only covers your mouth and nose; or a full-face, which covers your eyes as well (Figure 3).

Powered Air-Purifying Respirators (PAPRs) use a battery powered fan to draw air through the filter to the user. They are easier to breathe through—however, they need a fully charged battery to work properly. They use the same type of filters/cartridges as other air-purifying respirators. They may be attached to a full-face headpiece or a loose-fitting hood that fits completely over the user's head (Figure 3).

How Air-Purifying Respirators Work

Purifying elements for air-purifying respirators have cartridges that remove specific contaminants from the air passing through them (Figure 4). There are many different types of cartridges. The pesticide label specifies the specific cartridge you must use based on the anticipated hazard and vapor that will be present in your breathing zone.

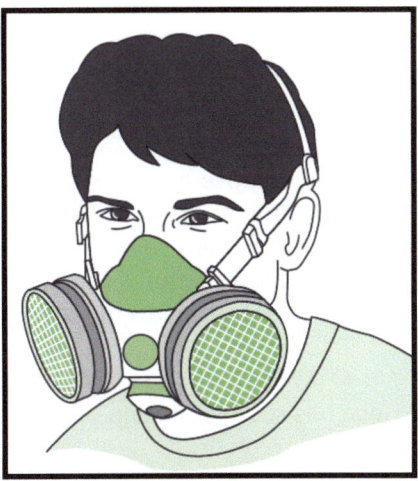 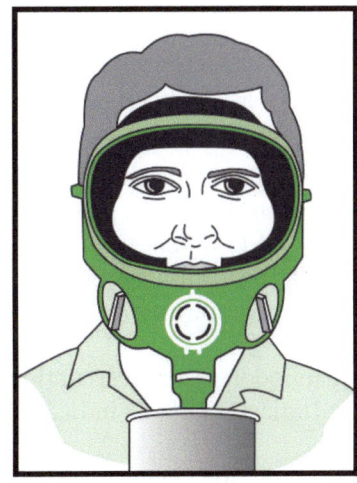 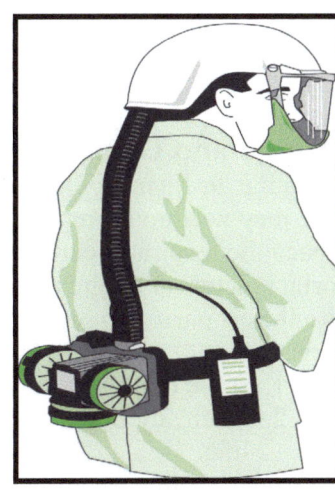

Figure 3: Half-face and full-face headpieces and Powered Air-Purifying Respirators (PAPR). These air purifying respirators use cartridges and/or metal canisters that contain material to filter the air the user breathes. The half-face headpiece (left) has dual canisters. The full-face headpiece (middle) has a single cartridge, although full face headpieces may also be equipped with dual canisters. Be aware that facial hair, earrings, scars, facial piercings, etc., may interfere with the seal of a tight-fitting headpiece. These will need to be addressed to ensure a tight respirator seal needed to effectively protect a worker's safety. A PAPR (right) is useful for people who may have problems breathing on their own through a regular air-purifying respirator. A full hood also helps people who cannot get a good seal on their face with tight-fitting headpieces. (Illustration © National Pesticide Applicator Certification Core Manual, NASDA).

The approved uses and limitations of different respirator cartridge types can be quickly and easily determined through color coding (Table 5-2, page 66). The effective life of respirator cartridges can vary depending on the type of fumigant used, the concentration of the fumigant in the space the respirator cartridges are used, the length of time the cartridges are used, humidity, and respiratory rate of the applicator. Each cartridge, and the instructional material it came with, will state its maximum limits. Cartridges also have an expiration date printed on them. Do not use a cartridge past its expiration date, even if it is still sealed and has not been used. You should crush expired canisters before discarding so they cannot be used.

When using a cartridge respirator, you must:
- Monitor the air concentration of the fumigant in structure or space you will enter to be sure the level does not exceed the level the canister is designed to purify.
- Make sure the cartridge you use in your respirator is appropriate for the fumigant used. Be familiar with the color coding scheme of cartridges (Table 5-2, page 66).
- Make sure the cartridge brand matches the model and manufacturer of the face piece. Cartridges and face pieces **are not** interchangeable among manufacturers.
- Only remove the cartridge seals when attaching it to the respirator. Be sure to remove both the top and bottom seal.
- Log the date and time when you removed the seals. This begins the life of the cartridge. Always dispose of any canister that has been unsealed for eight hours or more.
- Canisters and cartridges should be replaced whenever they are damaged,

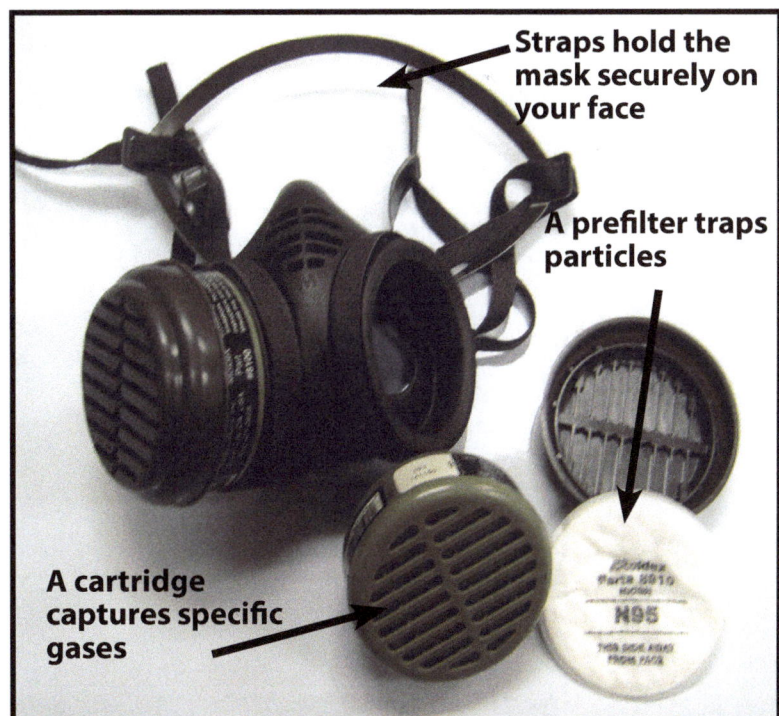

Figure 4: Air-purifying respirator cartridges. A typical air-purifying respirator has removable cartridges that filter out specific gases. Some cartridges also have a pre-filter to trap particulates. Cartridges are not intended to be permanently used and need to be replaced periodically. (Photo © UW-Madison Pesticide Applicator Training Program).

PERSONAL PROTECTIVE EQUIPMENT

soiled, or cause noticeably increased breathing resistance. Do not rely on odor or taste.
- Manufacturers may provide information to help determine the appropriate change out schedule for their products.

Cartridge and Canister Replacement

Because the air-filtering elements (cartridges and canisters) have a finite capacity to remove contaminants from the air, they need to be replaced periodically to keep you protected. The specific fumigant, its concentration in the space the respirator cartridges are used, and the length of time the cartridges are used are some of the most important factors affecting how long cartridges and canisters will last. California regulations require employers to ensure the air purifying elements, or the entire respirator (if disposable), are replaced according to the following criteria:

1. At the first indication of fumigant odor, taste, or irritation;
2. When any End of Service Life Indicator (ESLI) indicates the cartridge has reached its end of service life;
3. According to pesticide specific label directions;
4. According to pesticide specific directions from the respirator manufacturer; or
5. In the absence of pesticide specific instructions or indications, cartridges must be replaced at the end of each day's work period.

NOTE: Most respirator manufacturers will instruct users to change air-purifying elements at the "first indication of odor, taste, or irritation." This generic manufacturer statement should not be viewed as a direction regarding service life.

In most cases, pesticide specific cartridge change out information is not available. However, for a limited number of chemicals that have use in other industries (e.g., phosphine), this information may be available from the respirator manufacturer.

Table 5-2: Respirator Cartridge Color Chart.

Color Coding for NIOSH Approved Respirator Cartridges and Canisters	
ALWAYS Consult the Pesticide Label for Respirator and Cartridge Selection	
Used for:	**Assigned Color**
Organic Vapors	BLACK
Ammonia	BRIGHT GREEN
Acid Gasses	WHITE
Organic Vapors (OV) and Acid Gasses	YELLOW
Organic Vapors (OV), Ammonia, Acid Gasses	OLIVE/BROWN
High Efficiency (HE) Filter, P100 Filters	MAGENTA
Organic Vapors **AND** High Efficiency (HE) Filter, P100 Filters	BLACK / MAGENTA
Organic Vapors (OV), Acid Gasses **AND** High Efficiency (HE) Filter, P100 Filters	YELLOW / MAGENTA
Organic Vapors (OV), Ammonia Acid Gasses **AND** High Efficiency (HE) Filter, P100 Filters	OLIVE/BROWN / MAGENTA

Atmosphere-Supplying Respirators

Atmosphere-supplying respirators provide clean air from an independent source unlike cartridge respirators that purify the surrounding air. There are two types of atmosphere-supplying respirators: the self-contained breathing apparatus (SCBA), and the *supplied-air respirator* (SAR). Both have a full-face mask (or hood) that delivers air to the fumigator from a compressed air tank or from an ambient air pump. The main difference between these two is that with a SCBA, you carry the air with you in a tank on your back, while with a SAR, the air tank or pump is located outside the fumigation area.

SCBA: Because you carry your air, you do not need to be connected to a stationary source of air, which does not restrict movement and gives you mobility (Figure 5). However, the weight and bulk of a SCBA can make work strenuous and difficult. Do not confuse SCBA with SCUBA (self-contained underwater breathing apparatus). These systems are very different. You cannot interchange their uses.

Self-contained units provide a limited quantity of air. Once you exhaust the air supply, the system ceases to provide protection. A pressure demand regulator admits fresh air into the mask as you begin to inhale to conserve the air supply. Airflow diminishes when you exhale. Manufacturers equip these units with warning devices to alert users when the air supply is getting low.

Figure 5: Self-Contained Breathing Apparatus (SCBA). This type of atmosphere-supplying respirator supplies clean air from a tank you carry. They have an alarm to warn you when the air supply is low.(Illustration © National Pesticide Applicator Certification Core Manual, NASDA).

Cleaning and maintaining a self-contained supplied air respirator is critical to its safe operation. The mask must fit properly to seal on the face. Exhalation valves must be in good working order to prevent any outside air from entering. Regularly inspect the air hose and replace it if it becomes cracked or shows wear. Keep pressure regulators clean, dry, and protected from damage.

SAR: SARs do not require filters or cartridges to remove the gas because they provide an outside source of clean air from a stationary source through a long hose from a pump or large stationary air tank (Figure 6). However, the length of available hose restricts the range of the wearer's mobility. The maximum hose length is usually 300 feet. The air is supplied to a facepiece, helmet, hood, or a complete suit depending on the level of protection needed. However, there are drawbacks to SARs as well. The long hose can get kinks, be cut, or get damaged in a way that cuts off or contaminates your air supply. The long hose can also restrict movement. If the hose attaches to an air pump, you must locate the pump in an area where safe, fresh air is available.

These respirators use either a hood with a clear face piece, a helmet with a clear face piece, or a strap-on half or full-face piece. The hood, helmet, and full-face piece models provide eye protection. Hoods and helmets provide a continuous flow of fresh air around the entire head, whether inhaling or exhaling. Because of continuous positive air pressure, hoods and helmets do not require critical sealing around the face, so they work even if the user has a beard, long sideburns, and /or wears eyeglasses. Replace the hood if it has holes or tears. Regularly inspect the air hose and replace it if it becomes cracked or shows wear.

Training is critical for the use of any SCBA or SAR. Before using these devices, you must be thoroughly trained on how to properly inspect, maintain, *fit test*, and store the device. You should be able to determine that it is in good working order prior to use, has the adequate air supply needed for the job at hand, and fits properly to provide an adequate seal around your face. You should also participate in periodic training exercises on the use of these devices.

Examples of Label Respiratory Requirements

A fumigant label will specify which type of respirator to use or that must be available in case of emergency. For example, a product label may specifically require:

- "A supplied air respirator (NIOSH approval number prefix TC-19C)" or
- "A self-contained breathing apparatus (SCBA) (NIOSH approval number prefix TC-13F)"

Figure 6: Supplied-Air Respirators (SAR). This type of atmosphere-supplying respirator weighs less than a SCBA and provides longer use. (Illustration © National Pesticide Applicator Certification Core Manual, NASDA).

Respirator Medical Evaluation

California regulations require a medical evaluation of any employee who may have to wear a tight-fitting respirator (such as a half- or full-face respirator, or an SCBA) while handling pesticides. The evaluation must be performed by a medical professional to ensure that the employee is healthy and physically able to use the respirator. The product label may require **all handlers** to undergo a medical evaluation as well.

After a preliminary screening with a questionnaire, or an equivalent medical examination, a medical practitioner may determine that a more in depth examination is required. For example, a fumigant handler who, at screening, is suspected by the medical professional of having a heart condition may be required to undergo a complete physical examination before they are fit tested for a respirator. Handlers who need to wear respirators must be reexamined by a healthcare professional if their health status, respirator type, or use conditions change.

Respirator Fit Test

After being cleared by a qualified medical practitioner to wear a respirator, California regulations require you to be *fit tested* with the specific respirator you will use (Figure 7). A fit test verifies that a respirator is comfortable, and correctly fits and protects the user. You must be fit tested **before** the first time you use your respirator, and you need to be re-tested at least annually to confirm that your respirator continues to protect you. Fit test methods are either "qualitative" or "quantitative." A qualitative fit test relies on the user's senses to detect a test agent, such as a smelling a distinctive and pungent oil or involuntary coughing as a reaction to an irritant smoke. A quantitative fit test uses an instrument to measure the effectiveness of the respirator. Follow-up fit testing is also required under the following circumstances:

- The style of the face piece has changed,
- The respirator size, model, or brand has changed,
- There is a physical change in the person's face (e.g., weight change or dental work) that would affect fit,
- The respirator's fit is unacceptable,
- The user requests it, or
- Employer policy requires it.

Fit (or Seal) Check

Conduct a *fit check* (also called a seal check) before each use (Figure 8). This test helps you make sure your respirator forms a complete seal around your face. Be sure to clean your respirator after each use according to manufacturer instructions, inspect it regularly, and store it properly.

Figure 7: Respirator fit test. During a fit test, a user might be challenged with an irritating smoke, or a substance with a distinct odor, such as banana oil. If the user can detect the smell of the challenge substance, their respirator does not fit correctly. (Photo © Pacific Northwest Agricultural Safety and Health Center).

How to Get a Respirator Medical Evaluation

Ideally, medical evaluations are best conducted by occupational physicians. Complete and bring the OSHA respiratory medical questionnaire to your medical appointment. The questionnaire can be obtained from OSHA's respiratory protection website.

If you are unable to find a medical professional to perform an evaluation, contact OSHA at: 800-321-OSHA (6742). The OSHA website also lists offices in California.

When to Wear a Respirator and Which Type to Wear

The short answer to when and which type of respirator to wear is when and what the label tells you! The answer depends on the fumigant you are using, and the exposure limits listed on the labeling or in California regulations. Fumigant labels list fumigant exposure limits which are the maximum fumigant concentrations applicators can be exposed to without using respiratory protection. Workers must wear respiratory protection when fumigant concentrations are greater than those limits. Concentrations are expressed as parts per million (ppm). There are other occupational exposure limits (OELs), including the Cal/OSHA Permissible Exposure Limits (PELs) and the American Conference of Governmental Industrial Hygienist's Threshold Limit Values (TLVs). These occupational exposure values represent the maximum air concentration of a chemical considered negligibly hazardous for most people when exposed 8 hours a day, 40 hours a week over the course of a career. These OELs can usually be found in the SDS or on the Cal/OSHA website. The goal is to keep your exposure to chemicals below these limits to protect yourself from adverse health effects (see Table 5-3, page 71).

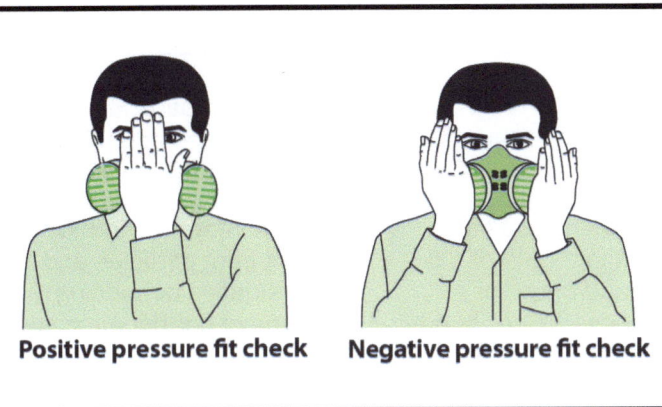

Positive pressure fit check   Negative pressure fit check

Figure 8: Fit (or Seal) check. For a positive pressure seal check, cover the exhalation valve on the front of the respirator and gently exhale. If you do not feel a rush of air around the faceplate, the seal is good. If you feel air leaking under the facepiece, reposition the facepiece and repeat the check until the seal is effective. For a negative pressure seal check, cover the inlet opening of each cartridge with your hands and inhale gently so the facepiece collapses. Hold your breath for about 10 seconds; if the facepiece stays collapsed, the seal is effective. If the facepiece expands, or air leaks under the facepiece, reposition it and repeat the check until the seal is effective. (Illustration © National Pesticide Applicator Certification Core Manual, NASDA).

Table 5-1. Respirator Types and Characteristics

Respirator by Air Source	Respirator Sub-Types	Other Names Used	Coverage	Characteristics
Air-Purifying Respirators	Filtering Facepiece	N-95, dust mask, dust/mist respirator	Typically mouth and nose.	Protects by filtration: Only protects against particulates, disposable, does not protect against fumigants.
	Elastomeric Half-mask	Gas mask	Mouth and nose.	Protects by filtration: Can protect from particles and vapors/gases depending on the type of cartridge/canister.
	Elastomeric Full-face	Gas mask	Mouth, nose, and eyes.	Protection by filtration: Can protect from particles and vapors/gases depending on the type of cartridge/canister.
	Powered air-purifying respirators	PAPR	Can have a tight fitting* half- or full-face mask, or a loose-fitting* hood that goes over the whole head.	Protection by filtration: Can protect from particles and vapors/gases depending on the type of cartridge/canister.
Atmosphere-Supplying Respirators	Self-Contained Breathing Apparatus	SCBA	Can have a tight fitting* half- or full-face mask, or a loose-fitting* hood that goes over the whole head.	Protection by supplying clean air in a tank carried by the wearer.
	Supplied-air respirator	Airline respirator	Can have a tight fitting* half- or full-face mask, or a loose-fitting* hood that goes over the whole head.	Protects by supplying the user with clean ambient air from outside the fumigated structure/area delivered through a hose connected to a compressed air tank.

* NOTE: Any "tight-fitting" head piece needs to be fit-tested. Loose-fitting pieces do not need fit testing.

TABLE 5-3. FUMIGANT CONCENTRATION REQUIRING RESPIRATORY PROTECTION:		
FUMIGANT	GAS CONCENTRATION (OEL)	RESPIRATORY EQUIPMENT REQUIRED
Phosphine	Less than 0.3 ppm	None required
	0.3 - 15 ppm	NIOSH-approved full face canister respirator – phosphine canister combination
	Greater than 15 ppm or when concentration is not known	NIOSH-approved self-contained breathing apparatus (SCBA)
Sulfuryl Fluoride	1 ppm or less	None required
	Greater than 1 ppm or when concentration is not known	NIOSH-approved self-contained breathing apparatus (SCBA)
Methyl Bromide	5 ppm or less	None required
	Greater than 5 ppm or when concentration is not known	NIOSH-approved self-contained breathing apparatus (SCBA)
Sulfur Dioxide	2 ppm or less	Non required
	Greater than 2 ppm	NIOSH-approved full-face respirator with an acid gas or combination organic vapor/acid gas cartridge
	Greater than 10 ppm or when concentration is not known	NIOSH-approved self-contained breathing apparatus (SCBA) or supplied-air respirator
Carbon Dioxide	Below 5,000 ppm	None required
	Between 5,000 - 30,000 ppm	For periods less than 15 minutes, no respirator required. For periods longer than 15 minutes a NIOSH-approved self-contained breathing apparatus (SCBA) or supplied-air respirator
	Greater than 30,000 ppm	NIOSH-approved self-contained breathing apparatus (SCBA) or supplied-air respirator
MITC	0.6 ppm or greater	NIOSH-approved air supplied self-contained breathing apparatus (SCBA) with organic vapor cartridge

Chapter 5 Review Questions

Correct answers are given on page 175.

1. Dry cotton gloves are required when applying _____.
    a. sulfuryl fluoride
    b. aluminum phosphide
    c. methyl bromide

2. Which respirator provides air from an outside source?
    a. air-purifying respirator
    b. canister respirator
    c. supplied-air respirator

3. A noticeable increase in breathing resistance when using a respirator is an indication that _____.
    a. respirator cartridge should be replaced
    b. respirator is too big for the user
    c. respirator is working effectively

4. When must a fumigant applicator be reexamined by a healthcare professional?
    a. if a new style of respirator will be used
    b. after respirator cartridges are replaced
    c. before each fumigation

5. Respirators that pass ambient air through an air-purifying element (filters, cartridges, or canisters) to remove contaminants from the air before it reaches the user are called _____.
    a. air-purifying respirators
    b. atmosphere-supplying respirators
    c. self-contained breathing apparatus

CHAPTER 6

Fumigation Methods and Application Equipment

LEARNING OBJECTIVES
- ☑ Describe the application methods and equipment commonly used for non-soil fumigation.
- ☑ Discuss the possible compatibility concerns for fumigants and application equipment, including tanks, hoses, and tubing.
- ☑ Describe how rate of air exchange, temperature, and sorption/desorption affect aeration time.
- ☑ Define half-loss time and load factor.
- ☑ Describe the basic procedures and precautions for aerating a fumigation chamber, a sealed building, or a tarpaulin fumigation.
- ☑ Discuss what to do when there is an equipment failure.

Terms to Know
The following are important terms to know from this chapter. They are explained and *italicized* in the text and defined in the glossary at the end of this manual.

Burrow Fumigation	Load Factor	Structural Fumigation
Chamber Fumigation	Quarantine Fumigation	Vertebrate
Half-loss Time	Spot Fumigation	

Steps to Fumigation

The objective of all non-soil fumigation methods is to introduce a lethal concentration of a gas into all parts of the treatment site (e.g., rodent burrow, wine barrel, warehouse, grain bin) and maintain that concentration long enough to kill all stages of the pests present. The site and/or items that need to be treated will usually dictate the best fumigation method for achieving pest control. Before we discuss fumigation methods in more detail, we need to discuss the basic steps that will be part of all (or nearly all) fumigations you might perform. These steps are to:
- Determine the appropriate fumigation method to use,
- Secure the fumigation site,
- Seal the site,
- Release the fumigant,
- Monitor the fumigant concentration,
- Determine exposure period,
- Aerate the site, and
- Clear the site.

In this chapter, we will focus on aerating the site. The other steps are discussed in succeeding chapters.

Fumigation Methods

There are different ways to categorize fumigation methods. Because of the diversity of structures, containers, commodities, products, and items that you might fumigate, some cross-over in categories is inevitable. For example, some may classify structural, chamber, and tarp as three distinct fumigation methods. However, if you enclose a structure (e.g., a grain mill) you are fumigating with a tarp, are you using a structural or tarp method? In this in-

stance, you are using both methods. The "method" you choose should be the one that makes the most sense for the situation at hand. The following is one way of categorizing fumigation methods:
- *Chamber*
- *Spot*
- *Structure*
- *Burrow*
- *Quarantine*
- Sewer root control

Chamber (or Vault) Fumigation

A fumigation chamber is a special type of vault designed specifically for fumigating items contained within it (Figure 1). In *chamber fumigation*, you can carefully control and monitor fumigation conditions. Use of a chamber can drastically reduce the amount of fumigant needed. The advantage of a fumigation chamber is to enable fumigations to be carried out efficiently, safely, and economically. Some basic elements are common in the design and construction of all fumigation chambers although variations can be incorporated to suit specific needs. An effective fumigation chamber is:
- Gas tight,
- Equipped with an efficient system for applying and distributing the fumigant,
- Equipped with an efficient system for removing fumigant at the end of a treatment,
- Situated to facilitate moving goods into and out of the chamber, and
- Designed so that it can be operated with minimal hazards to workers and the environment.

Figure 1: Chamber (or Vault) Fumigation. Just one of many different types and sizes of modern fumigation chambers. (Photos © USDA-ARS).

There are two main types of fumigation chambers: atmospheric and vacuum. An atmospheric chamber can be any airtight structure that is under ambient air pressure. Vacuum chambers are large steel structures where a vacuum can be applied. For many items, fumigation can be conducted at atmospheric pressure. However, vacuum fumigation is recommended for the treatment of certain densely packed or absorbent materials. It is also used when a rapid turnover of fumigated goods is required.

Spot Fumigation

Spot fumigation is the short-term fumigation of a site (e.g., machinery, wine barrels, utility poles, and small storage areas). For example, you can use it to control pests that infest whole foods and food particles that remain within processing equipment. You can use spot fumigation to control stored product pests in places such as food processing equipment (i.e., sifters, rollers, and dusters); related equipment in mills, food and feed processing plants, breweries; and similar industries.

Spot fumigation allows you to treat only areas where pests exist. This method saves time and money and puts less fumigant into the environment.

Equipment: For machinery and small storage areas you will need the same, or similar, sealing materials (tape, plastic sheeting, tarps, etc.) as you would for a *structural fumigation*. For wine barrel fumigation, you will need a bung (stopper) to seal the barrel. For utility poles, you will use a tight-fitting plug to seal the hole you drill into the pole. Depending on the item being fumigated, the seal may be treated wood (like a tight-fitting dowel or cork) or a plastic plug.

Tarp Fumigation: Tarp fumigation may be considered a form of spot treatment (or also a form of chamber fumigation where the "chamber" is made from a tarp). In tarp fumigation, you use tarps to treat small items rather than entire structures (Figure 2).

Place the items to be fumigated on a sufficiently airtight foundation, such as another tarp or on concrete. Cover the items with the tarp to ensure a tight seal. Seal all seams to create a gas tight enclosure. Provide support for the tarp over the items to be fumigated to allow room for gas expansion. Seal the edges of the tarp on the foundation by weighting them down with snakes filled with water or sand.

✋ Conduct tarp fumigations outdoors in the open, on loading docks, or in areas of buildings that allow safe aeration when the tarp is removed. You can also perform a tarp fumigation inside a building if you can assure it will not be occupied during fumigation and aeration.

Tarp Material: Tarps used in fumigation may be made of special materials (i.e., impregnated nylon) and/or be of a certain thickness (i.e., polyethylene sheeting of at least 4 mil thickness). Impregnated nylon tarps can be used several times because they are strong and resist ripping. You can clamp sections of impregnated nylon tarps together, so there is almost no limit to the size of the stack or structure that you can cover. Thinner polyethylene works better than fumigation tarps if folds are necessary (e.g., at corners). Polyethylene sheeting should not be reused since pinprick size or larger holes usually occur during handling. Refer to the label and applicator's manual for the type and thickness of tarp required.

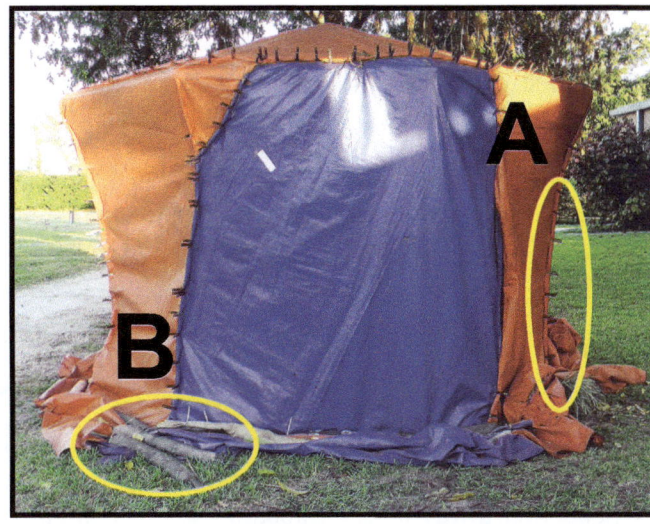

Figure 2: Tarp fumigation. Sometimes tarps are used to create a "chamber" to enclose items for fumigation. Note the clamps that hold and seal the edges of two tarps that come together (A), and the snakes that hold and seal the tarp edges to the ground (B). (Photo © Betsy Danielson, Iowa State University Extension and Outreach).

Fumigation of Structures

In California, the Structural Pest Control Board issues the license for the use of fumigants (e.g., sulfuryl fluoride) to control pests (e.g., termites, powderpost beetles, bedbugs, cockroaches, rodents, etc.) in structures (e.g. houses, buildings, or other similar structures). This study manual will not cover these uses. However, there are many items that are considered a "structure" but fall under DPR's regulatory scope. These include, for example, fumigations to control commodity storage pests in:

- A warehouse or factory.
- Grain bins, silos, or similar storage buildings.
- Boxcars, semi-trailers, or ships.

These structures can (and almost always need to) be sealed by tarps or tape-and-seal before introducing fumigant. These treatment sites will be discussed in more detail in later chapters.

Equipment

Sealing Materials: There are numerous sealing techniques, but the most commonly used supplies for sealing are plastic sheeting, adhesive tape, adhesive sprays, and expandable foam caulking. When using tarps, you will also need clamps to seal tarp edges, and water or sand filled snakes for ground seals.

Cylinders and Hoses: You often apply fumigant gases from a pressurized cylinder (Figure 3) directly into the treatment site through an introduction hose (also called a shooting tube). Sometimes fumigants get delivered as liquids, with the hose dripping the liquid into an evaporation pan. The rate of introduction of the fumigant is controlled

Figure 3: Releasing fumigant from a pressurized cylinder. When performing this task, always wear eye protection and make sure that the hose discharges away from you. (Photo © Betsy Danielson, Iowa State University Extension and Outreach).

by the length and inside diameter of the introduction hose. The longer the hose and narrower the inside diameter, the slower the introduction rate. When possible, it's a good idea to place the cylinders outside the fumigated space to reduce exposure for workers (Figure 4). For large sites, multiple hoses may be used to increase the introduction rate of a fumigant and more rapidly attain the effective concentration throughout the treatment area.

Introduction of Tablets or Pellets: To fumigate a building with aluminum or magnesium phosphide, place the required number of tablets or pellets (determined by volume of the space and the commodity) in each room of the building. You can place the tablets or pellets on a tray (e.g., cardboard tray) to contain the residue. **Never** place phosphide tablets or pellets on a wet surface or in standing water since it they will react with moisture and produce gas very quickly, and may possibly ignite or explode.

Figure 4: Releasing fumigant from outside the building. When possible, releasing the fumigant from outside the building is recommended as it increases safety by minimizing an applicator's exposure to the fumigant. This photograph illustrates a large fumigant operation involving several pressurized cylinders of fumigant located outside the structure with attached hoses to introduce the fumigant inside the structure to be fumigated. (Photo © Douglas Products).

For grain bins which are being loaded, pellets or tablets can be applied continuously by hand with an automatic dispenser on the belt, or into the fill openings. You can also use an automatic dispenser to add the fumigant into the grain stream in the up leg of an elevator.

When the grain bin is already loaded, a probe may be used to distribute tablets or pellets evenly throughout the treatment area (Figure 5). Insert the probe into the grain pile to the desired depth. Drop in the tablets or pellets. Pull the probe upward in evenly spaced intervals, stopping to drop in more product. Continue until the desired amount of product is dispensed. The last tablets or pellets applied through the probe should be about six inches below the surface. Repeat probing until the total amount of product for the fumigation is applied, distributing the probes evenly throughout the grain pile.

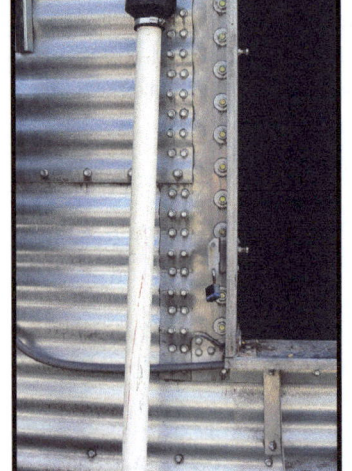

Figure 5: Phosphine Probes. You can buy commercially made phosphine probes or make them from electrical conduit, plastic pipe, or tubing (above). (Photo © Garo Goodrow, Penn State).

✋**IMPORTANT!** When probing, always wear a safety harness and rope along with your other PPE. Work in teams if possible.

Tarps: If you do apply the fumigant on the surface or with probes, after the application, you can cover the entire surface of the grain with polyethylene sheeting. Phosphine cannot easily penetrate polyethylene tarps. They will reduce gas loss substantially as well as reduce the amount of fumigant required since the fumigant cannot diffuse to the empty space above the grain.

Fans: You can use a fan to help introduce the fumigant when using cylinders and a hose (Figure 6). Regardless of whether you use a fumigant that comes in pressurized cylinders or as solid formulations, you may need multiple fans to spread the fumigant evenly and quickly throughout the space or commodity being treated. The fumigant must move into the structure's cracks and crevices, and into spaces within the stored commodity to contact all pests. Some grain storage structures have built-in aeration or recirculation systems that can aid in circulating and aerating the fumigant.

Figure 6: Fans. You can use fans to introduce fumigant to the site. The arrows in the photograph point to attached introduction hoses. Fans are also useful, and sometimes necessary, to facilitate aeration, especially under calm conditions. (Photo © Bonnie Rabe).

✋ When using methyl bromide, sulfuryl fluoride, or phosphine, you **must** use non-sparking fans. Heat sources (e.g., pilot lights, electric heating units, etc) can cause methyl bromide and sulfuryl fluoride to become corrosive and can also cause phosphine to ignite.

Monitoring Lines and Equipment: We discussed monitoring equipment in detail in Chapter 3, "Planning for a Fumigation". Just as you might need multiple introduction lines for a large treatment area, you might also need multiple monitoring lines. If the monitoring lines are long (several hundred feet), some labels recommend the use of a vacuum purge pump to ensure timely and accurate collection of air monitoring samples from all areas within the structure.

Burrow Fumigation

Fumigants may be used to control burrowing *vertebrate* pests. Vertebrates are animals with an internal backbone such as rodents (e.g., pocket gophers, ground squirrels). The Vertebrate Pest Control Research Advisory Committee (VPCRAC) Online Vertebrate Pest Control Handbook, the University of California Statewide IPM Program's Pest Notes Library, or the University of Nebraska's Prevention and Control of Wildlife Damage, are just a few resources on the correct identification and management of vertebrate pests.

At the time of this manual's writing, there are three types of fumigants registered by U.S. EPA and DPR for use in burrows: ignitable gas cartridges, carbon dioxide (dry ice pellets and in cylinders), and aluminum phosphide based tablets/pellets. Devices which generate carbon monoxide to control burrowing rodent pests are also available from several manufacturers. Note, carbon monoxide devices are not labeled as fumigants. These fumigants and devices are discussed in detail in Chapter 7.3, "Burrow Fumigation For Vertebrates."

Fumigants labeled for burrow use may have specific restrictions on how close to buildings they can be used. For example, aluminum phosphide labels will contain a statement such as:

> "FOR BURROWING RODENT APPLICATIONS: THE USE OF THIS PRODUCT IS STRICTLY PROHIBITED WITHIN 100 FEET OF ANY BUILDING WHERE HUMANS AND/OR DOMESTIC ANIMALS DO OR MAY RESIDE ON SINGLE AND MULTI-FAMILY RESIDENTIAL PROPERTIES AND NURSING HOMES, SCHOOLS (EXCEPT ATHLETIC FIELDS), DAYCARE FACILITIES AND HOSPITALS."

The use of burrow fumigants may also be restricted in some locations to protect non-target and/or endangered species. U.S. EPA's Bulletins Live! Two and DPR's PRESCRIBE database are important resources that you can consult prior to fumigating burrows to determine any applicable location restrictions.

Just like a building fumigation, you must conduct pre-fumigation tasks to assess whether a *burrow fumigation* is a possible control option to use, and if it is, how it can be conducted safely and effectively. Some of these tasks include surveying the site, assessing the burrow(s), and determining if there are any non-target species that may be impacted by the fumigation.

Equipment and Introduction of the Fumigant: The equipment used varies depending on the fumigant chosen and the characteristics of the target pest's burrow system. For example, with aluminum phosphide, if the pest species has an open burrow system, after inserting the required number of tablets or pellets, each treated entrance is packed with crumpled paper before shoveling soil to completely cover the paper and seal the opening. This will prevent soil from covering the tablets or pellets and slowing down their action. Rocks, clods of soil, cardboard, etc. may be used for this purpose. If the species has a closed burrow system (such as pocket gophers in some situations), a probe is used to identify the main runway and the required number of pellets are delivered using the probe. You then close the probe hole and create a tight seal using a clod of soil or a sod plug or use the heel of your shoe to push sod and/or soil over the surface opening. If the probe hole is more than one inch in diameter, place crumpled paper in the hole before closing it with soil or sod.

Quarantine and Pre-shipment Fumigation

Quarantine fumigation is a commodity fumigation ordered by an importation regulating agency to protect domestic agriculture and the environment from imported exotic pests. The three primary gases used for quarantine fumigation are methyl bromide, phosphine, and sulfuryl fluoride. Agricultural inspectors check incoming containers to determine which commodities must be fumigated before they can be released to the market. The presence of certain pests can require that the entire container be fumigated.

Each required fumigation is directed and supervised by an officer of the USDA-Animal and Plant Health Inspection Service (APHIS) Plant Protection and Quarantine (PPQ) program.

> **More Information on Quarantine Fumigation**
>
> Quarantine programs are run by the United States Department of Agriculture Animal and Plant Health Inspection Service (USDA – APHIS).
>
> The agency also has an official fumigation treatment manual for commodities. You can download it as a PDF.

Commodities being exported from the United States or to other states within the U.S. also may come under pre-shipment/quarantine requirements depending on their destination. "Pre-shipment" is used to describe goods that are held for a maximum of 21 days after fumigation before export.

Equipment: The equipment used for quarantine shipments is the same as other types of fumigations. However, the equipment used for a quarantine fumigation will depend on where the fumigation will take place. Quarantine fumigations can take place in semi-trailers, shipping containers, warehouses, separately tarped items, rail cars, or ships.

Sewer Root Control

When sewer pipes get clogged by tree roots, pesticides are sometimes used to clear the pipes. One pesticide that has been used for decades for sewer root control is a combination of metam sodium (a fumigant) and dichlobenil (a growth inhibitor). Metam sodium breaks down into the gas methyl isothiocyanate (MITC) which kills plant roots.

Equipment: Metam sodium products are introduced into pipes as a dry foam. Applicators use special foam generating equipment to produce the foam, which is then applied to the interior of the pipe. A hose the same length as the pipe to be treated is inserted and used to introduce the foam. The pipe is then filled with foam as the hose is retracted. The foam collapses over a period of an hour or more and will adhere to the pipe and root surfaces. Sewer root control is discussed in more detail in Chapter 7.7.

Aeration

Aeration procedures vary according to the fumigant and the type of installation and/or items being fumigated. Always read and follow label instructions for the fumigant and for the situation for which it is being used. Plan for every possibility because it is difficult to predict changes in wind speed and direction or other possible weather conditions. Aeration under calm, cold conditions may take several days. You must always be sure a site is properly aerated before you let people enter the site without wearing the proper PPE.

Factors Affecting Aeration Time

In addition to the characteristics of the fumigant itself, the rate of ventilation and aeration is affected by other important factors such as the load factor, rate of air exchange, and temperature.

Load Factor

The amount of fumigant sorbed by materials is referred to as the *load factor*. Sorbed fumigant is not available to act as a fumigant but must still be removed during aeration. Some commodities are much more sorptive than others just as some fumigants are more subject to sorption than others. The greater the sorptive capacity of the fumigant and commodity, the longer the desorption process and the greater the aeration time needed.

The greater the surface area of the commodity, the greater the sorption rate and the longer the aeration period. For ex-

Figure 7: Using fans to increase cross ventilation. Fans are useful for aerating after a fumigation. Remember to use non-sparking fans to minimize the risk of fire or explosion. (Photo © Cardinal Products).

ample, open machinery aerates more rapidly than bagged flour. Make sure you factor in retention of fumigant by highly sorptive materials like grain, flour, meals, and jute bags.

Rate of Air Exchange
The rate of air exchange within the fumigated space is the most important factor affecting aeration. The exchange rate will be proportional to wind velocity through the area, size and arrangement of the fumigated site, and the mixing of gases. The time that it takes the initial concentration of fumigant to be reduced by half is referred to as *half-loss time* (HLT). In atmospheric chambers, an exchange time of one air change per minute is desirable. In other areas, the most effective and practical method to increase the rate of air exchange is to increase cross ventilation using fans (Figure 7). Loaded areas aerate much more slowly than empty areas. The rate of fumigant lost from an enclosure is also due to leakage and fumigant sorption.

Temperature
Higher temperatures favor diffusion of the fumigant and increase the rate of desorption. Longer aeration times are often required when aerating unheated or refrigerated commodities.

Aeration Procedures
Aeration releases free fumigant gas and aerates commodities in the fumigated space immediately following fumigation. Procedures for aeration and ventilation vary with the fumigant and the structure and/or commodity being fumigated. Have an aeration plan before starting the fumigation, including having available backup fans and alternative sources of electricity such as a generator.

Fumigation Chamber Aeration
Aeration procedures for fumigation chambers vary depending on whether the chamber is inside or outside a structure. When a fumigation chamber is inside a building where employees are likely to be present, provide intake and exhaust stacks. The exhaust stacks must lead outside the building. Open the intake and exhaust stacks after the fumigation is completed. You can use the air circulation equipment in the chamber to move air from inside the chamber to outside.

When a chamber is outside, you may aerate it safely by opening the door slightly at the beginning of the aeration period and turning the blower on. Prop the door partially open so it cannot accidentally close. If the door should accidentally close, the partial vacuum created by the blower may damage the chamber. Vent air discharged from the blower to outside of the chamber.

No one should be near the door or the exhaust when the blower is turned on. You can fully open the doors after about 15 minutes, but do not let workers enter the chamber until safe air concentrations have been confirmed with an appropriate monitoring device.

Building Aeration
When aerating a fumigated building such as a mill, food processing plant or warehouse for storage pests, open ground floor vents, windows, and doors from the outside whenever possible. Aerate the building until air monitoring indicates that it is safe to enter the building. If you must enter before the fumigant concentration has decreased to a safe level, wear the required and specified respirator, generally a SCBA.

Entering Building Before Complete Aeration
If you must enter a fumigated space or structure before aeration is complete, do so for short periods of time. You should leave the structure every 15 minutes or so for fresh air. At least two trained and certified applicators must work together. Both applicators must wear the respirator type specified on the label (generally a SCBA). Follow label instructions on opening the structure as these instructions are intended to minimize fumigant exposure to handlers and bystanders. Open ground floor vents, windows, and doors first, particularly those that were not previously opened from the outside. Open only those windows that will provide thorough cross ventilation. Work upward through the building, one floor at a time, opening just enough windows to provide the needed ventilation. Also turn on any fans that might assist in aeration. Allow only properly trained and equipped handlers into the building until air monitoring indicates the fumigant concentration is below the threshold level in all parts of the building.

Tarpaulin Aeration

Tarpaulin aeration procedures also vary slightly based on where the fumigation occurred. When aerating loads under tarps in buildings or outside on still, humid days, make an opening by lifting the tarp on the end opposite the blower or source of air and discharge the fumigant with the blower. If blowers or strong cross ventilation are not used, lift the tarp at the corners and raise it to allow the fumigant concentration to decrease. Remove the tarp when air monitoring indicates aeration is complete.

If aeration is in the open and a breeze is blowing, first lift the end or side of the tarp opposite the direction of wind movement, then safely open the portion of the tarp on the windward side.

Equipment Compatibility and Equipment Failure Concerns

Make sure you use the proper equipment for whatever fumigant you will be using. The label and applicator manual will tell you if there are materials you should not use with a particular fumigant product. The chemical reactivity and flammability of various fumigants were discussed in Chapter 1. The discussion also included examples of materials that should not be exposed to specific fumigants.

Equipment Failure

Applicators should always be prepared for the possibility of equipment failures and other emergencies. In fact, planning for emergencies should be built into your Fumigation Management Plan (FMP), as we discussed in Chapter 4, "Fumigation Safety."

Applicators should always be prepared for the possibility of equipment failures and other emergencies. Planning for emergencies should be part of planning for any fumigation, regardless of whether the label requires a Fumigation Management Plan (FMP).

Knowing ahead of time how to react to certain equipment failure issues will help you address them safely and effectively. If possible, you should rehearse the fumigation process, responding to common equipment failure issues, and emergency evacuation procedures. Rehearsals provide the opportunity to not only detect problems in your emergency plans but to also correct them.

> If an equipment failure (e.g., an introduction hose comes loose from a tank, a tarp seal fails, etc.) results in fumigant being released where it shouldn't, always make sure to wear the proper PPE to protect yourself before attempting to make repairs or help a coworker who may have been accidentally exposed to the fumigant.

You should, whenever possible, have backup equipment and/or replacement parts for application equipment, PPE, hoses, clamps, sealing material, gas detectors, and other equipment. In some cases, this is mandated on labels. For example, if a label calls for a SCBA, you must have a second SCBA on hand during each fumigation in case you need to make an emergency entrance into a fumigation enclosure. The second SCBA is for potential use by another trained person who must be on site if you enter the fumigation enclosure.

California regulations require employers to have an accident response plan at the fumigation worksite to provide instructions to protect employees during situations such as spills, fire, and leaks. The accident response plan should include two core pieces of information. First, the plan should include information regarding how to secure the area where the accident occurred. Second, the plan should provide the contact information of key people including, but not limited to, the operator of the property, the local fire department, the health department, the hazardous materials response team, and poison control.

> **REMEMBER!** For products that require a Fumigation Management Plan (FMP), you must have an available and easily accessible list, including phone numbers, of the people, civil agencies, and emergency services to contact in case of an emergency.

Chapter 6 Review Questions

Correct answers are given on page 175.

1. The amount of fumigant sorbed by the materials being fumigated is called _____.
    a. half-loss time
    b. load factor
    c. fumigant coefficient

2. The short-term fumigant treatment to only certain areas where pests exist in a site is known as _____.
    a. spot fumigation
    b. temporary fumigation
    c. vacuum fumigation

3. If tarp aeration is outside and a breeze is blowing, the first part of the tarp that should be lifted is the _____.
    a. portion on the windward side
    b. end or side opposite the direction of wind movement
    c. northernmost edge or corner

4. Which of the following is a type of fumigation chamber?
    a. vacuum
    b. hydraulic
    c. industrial

5. Which of the following fumigants is registered by the United States Environmental Protection Agency and the Department of Pesticide Regulation for use in rodent burrows?
    a. carbon dioxide
    b. sulfuryl fluoride
    c. metam sodium

CHAPTER 7.1

Chamber Fumigation

LEARNING OBJECTIVES
- ☑ Describe the uses of chamber fumigation.
- ☑ List any specialized equipment needed for a chamber fumigation.
- ☑ List potential pests you can control through chamber fumigation.
- ☑ Explain any unique label requirements for chamber fumigation.

Terms to Know
The following are important terms to know from this chapter. They are explained and *italicized* in the text and defined in the glossary at the end of this manual.

Atmospheric Chamber
Chamber Fumigation
Restored Pressure Fumigation
Sustained-Vacuum Fumigation
Vacuum Chamber

Fumigation Chambers

Fumigation chambers (sometimes called fumigation vaults) are well sealed structures that may be located inside or outside main buildings. Some chambers are specially built for fumigation, while others are modified rooms or buildings. *Chamber fumigation* is useful because you can carefully control and monitor the fumigation conditions. Chambers can also significantly reduce the amount of fumigant you need. While well sealed buildings, grain bins, boxcars, or tarped constructions can be considered "chamber" fumigations because they are all enclosed structures, this chapter focuses on structures specifically designed for fumigation.

Chamber fumigation can be used for many items and commodities including fresh produce, packaged foods, bagged or baled agricultural products, museum specimens, furniture, high value garments, and similar items.

> **Museum Fumigations**
>
> Museums store a wide variety of valuable and vulnerable material (i.e., wooden artifacts, wool clothing, books, papers and parchments, dried plant specimens, and prepared mammal and bird specimens) that can be destroyed by pests. These pests of stored product, fabric, wood, and paper include wood-destroying insects, clothes moths, carpet beetles, cock- roaches, silverfish, cigarette and drugstore beetles, and dermestid hide beetles. Most museums have programs in place to reduce the risk of introducing pests, including the fumigation of all material brought into the museum. These are usually conducted via chamber fumigation.

Pests Controlled
Because the range of items that can be fumigated in chambers is so broad, so too, is the range of possible pests. Quarantined food and other items being imported or exported may contain any number of pests including food and grain destroying moths and beetles, wood destroying insects, and rodents.

Purpose of Chamber Fumigation
The purpose of a fumigation chamber is to allow fumigations to be carried out efficiently, safely, and economically. The design and construction of fumigation chambers include basic design elements with variations to suit specific needs. Some of these elements include that the fumigation chamber is:
- Gas tight,
- Equipped with an efficient system for applying and distributing the fumigant,
- Equipped with an efficient system for removing fumigant at the end of a treatment,
- Situated to facilitate moving goods into and out of the chamber, and

- Designed to minimize hazards to workers and the environment.

Fumigation Practices

Many of the practices involved in preparing for a chamber fumigation are no different than those already discussed. Refer to Table 4-1: Fumigant Practices in Non-Soil Fumigation (Chapter 4, page 57) for practices applicable to chamber fumigations.

Advantages of Chamber Fumigation

The advantages of using fumigant chambers are:
- You do not have to calculate the volume of the chamber each time you use it because it is constant.
- You often can use significantly less fumigant compared to other methods.
- Special preparation of the commodity is not necessary (i.e., restacking, reorganizing, moving commodities).
- Almost any fumigant can be used to treat a variety of items.
- You can monitor fumigant concentrations within the chamber through a permanent system.
- You can conduct fumigations more safely than in other structures (e.g., boxcars, ships) or under tarps because the chamber is specifically designed to properly deliver, contain, and exhaust fumigants,
- Chambers can be heated for cold weather fumigations.

Disadvantages of Chamber Fumigation

The disadvantages of fumigation chambers are:
- The costs of setting up the chamber.
- Moving commodities to and from the chamber.
- The limited capacity of a fumigation chamber.

Types of Fumigation Chambers

There are two basic types of fumigation chambers:
- *Atmospheric chambers.*
- *Vacuum chambers.*

Atmospheric Chambers

An atmospheric chamber is a structure under normal air pressure. Some are specially built for fumigation while others are modified from existing structures. You can construct a suitable, low-cost atmospheric chamber using a gas tight room with an appropriate door. The chamber should be equipped with features to apply, distribute, and remove a gas. You may add heating with the appropriate precautions. Position the chamber in a location that will allow you to easily move items/commodities in and out. Also, be sure to minimize hazards to workers and the environment. Atmospheric chambers should not be within or connected to other structures where fumigants may contaminate the air.

Vacuum Chambers

In vacuum fumigation, most of the air in the chamber is removed before the fumigant is introduced. It is necessary to have a specially constructed chamber, usually made of steel, that can sustain a vacuum. This chamber should also be equipped with fans or recirculating systems to ensure even distribution of the fumigant and the necessary pumps, valves, and pipes for introducing and exhausting the fumigant.

The vacuum denies the insect of oxygen and facilitates rapid penetration of the fumigant into the commodity. As a result, a vacuum fumigation treatment may be completed in 1½ to 4 hours compared to fumigations at atmospheric pressure that can take 12 to 24 hours to complete. This shortened fumigation period is desirable when a quick turnover of fumigated goods is required, as in a busy seaport or large warehousing operation.

Vacuum fumigation can be used to treat densely packed items and other materials that are difficult to penetrate at atmospheric pressures. It is also useful for plant quarantine and products that need to be treated quickly.

Not all fumigants are compatible with vacuum fumigation. Phosphine, for example, can explode under vacuum conditions (Figure 1). Always read the label to be sure the product is safe to use in a vacuum chamber. Some items and commodities (e.g., fresh fruits and vegetables) cannot withstand the reduced pressure of a vacuum chamber.

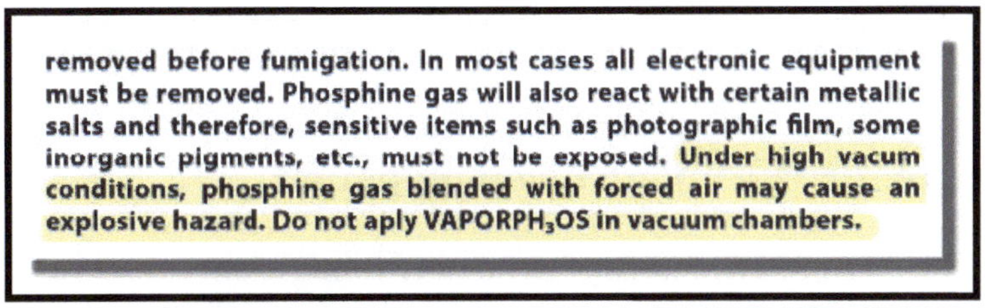

Types of Vacuum Fumigations

There are two types of vacuum fumigation: *sustained-vacuum fumigation* and *restored pressure fumigation*.

> removed before fumigation. In most cases all electronic equipment must be removed. Phosphine gas will also react with certain metallic salts and therefore, sensitive items such as photographic film, some inorganic pigments, etc., must not be exposed. **Under high vacuum conditions, phosphine gas blended with forced air may cause an explosive hazard. Do not aply VAPORPH$_3$OS in vacuum chambers.**

Figure 1: Label warning on the use of vacuum chambers. The label portion above warns users of the danger of using a phosphine gas product in a vacuum chamber.

The sustained-vacuum method starts when you reduce the pressure inside the chamber and introduce the fumigant. The slightly reduced pressure (vacuum) is held until the end of the treatment.

In the restored-pressure method, you lower the pressure, introduce the fumigant, and then restore the pressure in one of four ways:

1. Gradual Restoration: Release the fumigant and then slowly introduce air until the air pressure returns to normal. This usually takes 2 to 3 hours.
2. Delayed Restoration: Hold the vacuum for about 45 minutes following discharge of the fumigant. Then, allow air to rapidly enter the chamber.
3. Immediate Restoration: Just after releasing the fumigant, rapidly let air into the chamber by opening one or more valves.
4. Simultaneous Introduction of Air and Fumigant: Use special metering equipment to release a mixture of air and fumigant into the chamber.

Other Fumigation Considerations

Review the following chapters for key concepts applicable to chamber fumigations:
- Choosing the proper fumigant: Chapter 1, "Fumigant Basics."
- Site characteristics that influence a fumigation, calculating fumigant dosages, air monitoring equipment: Chapter 3, "Planning for a Fumigation."
- Fumigant Management Plans: Chapter 4, "Fumigation Safety."
- Securing the site, warning signs, protecting applicators, and others: Chapter 4, "Fumigation Safety," and Chapter 5, "Personal Protective Equipment."
- Application equipment and aerating the site: Chapter 6, "Fumigation Methods and Application Equipment."

Chapter 7.1 Review Questions

Correct answers are given on page 175.

1. What is a disadvantage of a chamber fumigation?
    a. fumigants can diffuse through outside walls of the chamber
    b. chambers cannot be heated
    c. it can be difficult to move commodities to and from the chamber

2. A fumigant that can explode when used in vacuum chamber fumigations is _____.
    a. phosphine
    b. methyl bromide
    c. sulfuryl fluoride

3. Which of the following is a type of vacuum fumigation?
    a. sustained-vacuum
    b. quarantine-vacuum
    c. atmospheric-vacuum

4. The type of chamber fumigation with a shortened fumigation period resulting in a quick turnover of fumigated commodity is _____.
    a. atmospheric chamber fumigation
    b. vacuum chamber fumigation
    c. hyperbaric chamber fumigation

5. Which of the following is an advantage of chamber fumigation?
    a. aeration is not needed or required
    b. each chamber is designed to treat a specific item
    c. special preparation of a commodity is not necessary

CHAPTER 7.2
Commodity, Post-Harvest, and Quarantine Fumigation

LEARNING OBJECTIVES
- ☑ Describe the uses of commodity and post-harvest fumigation.
- ☑ List any specialized equipment you need.
- ☑ Briefly outline potential pests you can control.
- ☑ Explain any unique label requirements for commodity and post-harvest fumigation.

In this chapter we will discuss fumigating structures that contain food, feed, or other commodities and/or the commodities themselves. There are laws and/or regulations governing "quarantine fumigations" or fumigation of certain commodities to prevent movement of agricultural pests into or within the United States. The methods and requirements of commodity, post-harvest, and quarantine fumigations are the same as, or similar to, other use patterns discussed in this study manual. They will apply regardless of whether the fumigation is performed under a tarp, or within a fumigation chamber. In this chapter we will focus on fumigation of:
- Farm structures containing commodities, such as grain bins, silos, barns or other storage structures, and
- Commodities not enclosed in any structure.

Commodity, post-harvest, and quarantine fumigations are designed to rid areas of pests that are damaging or could cause damage to food, feed, or other commodities.

Pests Controlled
The range of possible pests for commodity, post-harvest, and quarantine fumigation is vast, but typically they include food and grain destroying insects and possibly rodents.

Purpose of Commodity Fumigation
Commodities are fumigated to eradicate pests as they are moved between producers, warehouses, and retail markets. The principles of commodity fumigation are the same as those for fumigating structures: you use or create a confined air space around a commodity where you can release a fumigant to eliminate pests infesting the commodity.

Quarantine fumigation is simply a commodity fumigation required by an importation regulating agency to protect domestic agriculture and the environment from imported exotic pests.

Fumigants Used
The primary fumigants used for commodity fumigation are methyl bromide, aluminum and magnesium phosphide, liquid phosphine, sulfuryl fluoride, and carbon dioxide. Sulfur dioxide may also have some uses in certain situations.

Fumigation Practices
Many of the practices involved in preparing for a commodity, post-harvest, or quarantine fumigation are no different than those already discussed. Refer to Table 4-1: Fumigant Practices in Non-Soil Fumigation (Chapter 4, page 57) for practices applicable to commodity, post-harvest, and quarantine fumigations.

Other Fumigation Considerations
Review the following chapters for key concepts applicable to the fumigations in this chapter:
- Choosing the proper fumigant: Chapter 1, "Fumigant Basics."
- Site characteristics that influence a fumigation, calculating fumigant dosages, air monitoring equipment: Chapter 3, "Planning for a Fumigation."

- Fumigant Management Plans: Chapter 4, "Fumigation Safety."
- Securing the site, warning signs, protecting applicators, and the public: Chapter 4, "Fumigation Safety," and Chapter 5, "Personal Protective Equipment."
- Application equipment and aerating the site: Chapter 6, "Fumigation Methods and Application Equipment."

Commodity and Post-Harvest Fumigations

These include the fumigation of structures such as grain bins, silos, barns, or other storage structures which contain harvested food or feed commodities, as well as stacks or piles of commodities (food or feed commodities and non-food commodities like wood or timber) not enclosed in any structure.

Grain Fumigations

Corn, wheat, oats, and other grains make up the bulk of commodities that are fumigated. Most fumigants kill all insect life stages when used correctly. Fumigant molecules diffuse through the spaces between grain kernels as well as into the kernels themselves. Fumigants can penetrate places that are inaccessible to insecticide sprays or dusts.

Factors Affecting Grain Fumigation

There are several factors you should consider to ensure that a fumigation will be both safe and effective.

Insect Pests Considerations

There are a great number of insect species that will feed on and damage grain. Although a fumigant can kill all insect life stages, it may not provide adequate control if the grain to be fumigated is crusted, caked, or is high in moisture. These conditions must be corrected before a fumigation to ensure efficacy (Figure 1).

The goal of fumigation is to introduce a specified concentration of gas into all parts of the grain mass and to maintain that concentration long enough to kill all stages of insects present. Keep in mind, however, that once it diffuses out of the grain, the fumigant

Figure 1: Insect webbing on stored grain. Indian meal moths can leave a large amount of webbing (silk) on top of stored grain as shown in this picture (shiny material). Webbing, like crusted grain, can inhibit the movement of fumigant into the grain. (Photo © Phil Sloderbeck, Kansas State University, Bugwood.org).

does not provide residual protection and the grain is susceptible to reinfestation. Grains harvested and stored in the summer when temperatures are hot are more susceptible to pests than grains harvested in the fall when temperatures are cooler. As a rule, the longer a commodity is stored at temperatures between 60°F to 90°F, the greater the chance of pest problems.

Temperature and Moisture Effects

Temperature controls the speed of penetration and release of the fumigant into and out of grain. Low grain temperatures (less than 50°F) can slow down gas movement enough to result in inadequate control. You may need higher dosages, longer fumigation times, and prolonged aeration when temperatures are moderately low (less than 60°F). It is best to fumigate when grain temperature is above 70°F. Adjustments for grain temperature are given on the label.

Temperature also affects how easily insect pests are killed. In general, insects are more difficult to control at lower temperatures. Many stored commodity insects are dormant below 50°F. Because these insects breath at a very slow rate during dormancy, they may not take in enough fumigant to kill them, or the exposure time needed to kill them would have to be so long it would be prohibitive. In general, insects develop (respire, feed, grow) more rapidly as the temperature increases up to a maximum that varies by insect species. Although many label requirements state a minimum of 40°F to fumigate, fumigating at temperatures of 60°F or higher is generally more effective for insect control.

High moisture and dockage (easily removed waste material in wheat and other grains such as chaff, broken kernels, etc.) in grain slows the movement of the fumigant. The gas is absorbed and held by moist or finely divided pieces of grain and can result in available fumigant gas concentrations that are below what is required to kill insects. The moisture content of grain also influences the penetration of fumigant gases by altering the rate of sorption. In general, moist grain requires an increase in dosage or an extended exposure period to compensate for the reduced penetration by, and increased sorption of, the fumigant.

Storage Structures

Grain that might need fumigation is typically stored in large bins. Grains may also be stored in vehicles, such as boxcars or ships (refer to Chapter 7.6 for the discussion on fumigating vehicles). Since most storage structures are not completely airtight, they almost always require sealing. Grain bins are not airtight by design to prevent grain from spoiling. Fumigated grain bins are typically sealed using the tape-and-seal method. Sheeting is placed directly over the grain inside the bin and a tight seal is made by tucking the plastic under the grain at the edges. This also reduces the amount of fumigant needed, since you are not fumigating the empty air space above the level of the grain. Aeration systems, bin eaves, etc., are taped and sealed to prevent leakages as well.

Some storage buildings (i.e., wooden storage buildings) will require you to use tarps over the entire structure. Although there may be label recommendations for fumigation of grain in wooden bins, fumigating these structures can be costly due to poor control of fumigant concentration resulting in a higher dosage of the fumigant needed to reach the effective concentration.

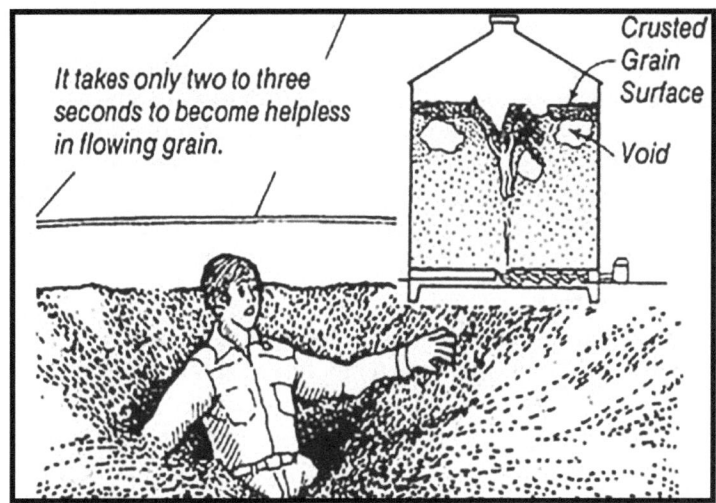

Figure 2: A hazard of entering a grain bin. Grains in grain bins can sometimes form crusts on top that feel stable but hide voids below where you could break through. There are other dangers as well. Never inspect grain bins alone. Use a harness and tether or other safety device anytime you work in a grain bin. (Illustration © U.S. Occupational Safety and Health Administration).

Sealing

Any structure you fumigate needs to be sealed before introducing a fumigant. The importance of sealing cannot be overemphasized! Fumigant will leak from a poorly sealed structure, resulting in a failed fumigation and a health hazard to people outside the structure. Be certain to close all doors and windows, seal vents and other openings, and repair cracks in the structure, including broken windowpanes. Take special care to seal off adjacent storage or work areas that are not to be fumigated.

Grain Bin Safety

You may need to enter a grain bin either before the fumigation to inspect it, during the fumigation to apply the fumigant, or afterwards to aerate and/or evaluate the job. Grain bins can be very dangerous places (Figure 2). Many people have been covered by grain and have suffocated. According to Purdue University, there was an average of 35 reported grain handling incidents per year between 2005 to 2015, of which 60% to 70% were fatal. A team inspection can minimize some dangers. One member of the team should wear a safety line while inside the bin with the second person handling the line on the outside. In addition, place a sign outside the bin to warn others that someone is inside the grain bin.

Fumigating Grain with Phosphine

Phosphine producing fumigants (aluminum and magnesium phosphide) are among the predominant fumigants used for the treatment of bulk stored grain throughout the world.

You can apply phosphine producing pellets or tablets on the conveyor belt or into the fill openings as the grain is loaded into the bin by hand or by an automatic dispenser. You can also use an automatic dispenser to add the tablets or pellets into the grain stream in the up leg of an elevator.

If it will take more than 24 hours to fill a bin, you should not fumigate the grain by continual addition of aluminum phosphide into the grain stream. Instead, fumigate the bin once it is filled. Make sure to lengthen fumigation periods to allow for diffusion of gas to all parts of the bin if the pellets or tablets have not been applied uniformly throughout the grain mass.

Ensure Even Distribution
Evenly distribute the aluminum or magnesium phosphide tablets or pellets on the surface of the grains. You can also use probes to introduce the tablets or pellets deeper into the grain. Evenly space the probes to ensure even distribution of the tablets/pellets. Insert the probe into the grain and when it is approximately 5 feet into the grain, place the first tablet or pellet in the probe. Raise the probe 1 foot and add the next tablet or pellet. Repeat this procedure until the probe is at about 6 inches from the surface and add the last tablet or pellet.

Recirculation
Some grain storage structures have built-in aeration or recirculation systems to regulate the temperature and moisture content of the grain. During fumigation, you can sometimes use these systems to distribute fumigants throughout the grain mass. However, be aware that not all storage structures will have these systems.

Figure 3: Recirculation. Using a recirculation system when fumigating a grain bin or other storage structure helps the quick and uniform penetration of phosphine throughout the commodity. In some cases, recirculation can reduce the needed dosage. (Photo © Jeffrey Jones).

It is often desirable to use a low flow recirculation system for phosphine gas in certain bulk storages (Figure 3). Recirculation usually involves the application of a fumigant to the surface of the commodity. The phosphine gas is then continuously or intermittently drawn out of the over space and blown into the bottom of the storage using specially designed low volume fans and duct work.

Stack Fumigations
Sometimes, it is more feasible to simply cover the commodity instead of the whole structure. In this instance, you must seal the commodity using tarps or other gas impermeable sheeting to create a temporary chamber. These are sometimes called "stack fumigations" (Figure 4).

You need to place the items on a sufficiently airtight foundation (i.e., on another tarp, or on concrete and covered

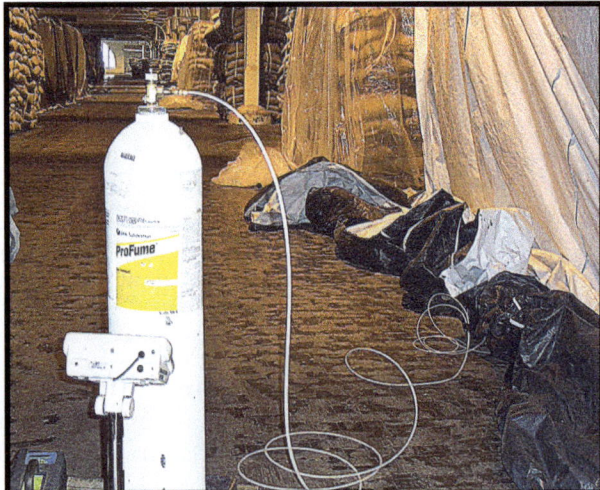

Figure 4: Stack fumigations. Stacked pallets of bagged cocoa beans are fumigated in a warehouse (right) and timber being fumigated outdoors under plastic sheeting (left). Note that the timber is also underlain by plastic sheeting to prevent the fumigant from dissipating into the soil. (Photo credits: (top) © Carl Schnabel; (bottom) © Douglas Products).

with a fumigation tarp) to ensure a gas-tight seal. Support the tarp over the items to create a gas expansion dome. Seal the bottom edges of the tarp with loose sand or with snakes.

Indoor Stack Fumigations
Indoor stack fumigation helps protect from wind and rain. However, most indoor treatments require you to evacuate the entire building, post warning signs, and monitor the area regularly. Some fumigant labels may allow work to continue in other parts of the building as long as the treatment area remains clear. If the commodity to be treated is in an unsuitable indoor site (e.g., close to a frequently occupied office space), it may be better to move the commodity to a more suitable indoor location. Make this decision during your initial inspection of the structure. Place all commodities to be fumigated on pallets for ease of movement.

Ventilating Indoor Stacks
A well-ventilated site is required for exhausting gas when the tarp is removed from the stack. Most warehouses have high ceilings and several windows/doors that can be used for ventilation. Some gas will escape from the tarp even in the best conditions. Avoid areas where strong drafts are likely to occur. In warehouses, you must provide an exhaust system to the outside of the building. You must ensure that the exhausted gas does not reenter the building or endanger people working outdoors.

Outdoor Stack Fumigations
If you will be fumigating stacks outside of buildings, select a site that is semi-sheltered and offers some protection from wind. This may be the side of a warehouse or a building. Placing sand snakes or sandbags over the tarp can help to protect it against wind. You also need to plan for bad weather. If you know it will be stormy or windy, delay fumigation.

Outdoor stack fumigation may require stronger and thicker tarps. Polyethylene tarps should be at least 4 mils thick, but 6 mil tarps are preferred. The color of the tarp also makes a difference. Clear polyethylene tarps tend to become brittle from ultraviolet rays of the sun. There are several challenges with outdoor fumigation. It is more difficult to obtain a good ground seal outdoors. Additional work may be required to lay a tarp on the ground and set the stacked commodities on top of it to ensure a good seal. You should place braces so that rain does not accumulate in any low spot. These braces should be placed over the commodity but under the tarp.

Quarantine Fumigation
Quarantine fumigation is commodity fumigation which is required by an importation regulating agency to protect domestic agriculture and native habitats from imported exotic plant pests. It does not refer to a specific fumigation method. A quarantine fumigation can take place in a structure (e.g., warehouse), under stacked tarps, in fumigation chambers, in boxcars, ships, and more.

The Plant Protection and Quarantine (PPQ) division of the USDA Animal and Plant Health Inspection Service (APHIS) regulates the importation of plants and plant products in the U.S. USDA-APHIS has a treatment manual that outlines the procedures for treating commodities for quarantine shipment, including non-fumigant treatments (Figure 5). APHIS-PPQ regulates quarantine shipments entering the U.S. or sometimes moving from one part of the country to another. They do not regulate quarantine fumigations of plant and plant products from the U.S. and going to other countries.

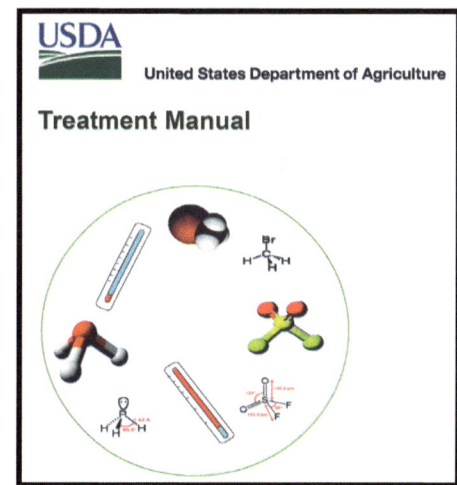

Figure 5: The USDA Treatment Manual for quarantine treatments.

Fumigants Used for Quarantine Shipments
At the time of this manual's writing, APHIS-PPQ allows the use of sulfuryl fluoride, methyl bromide, and some formulations of aluminum and magnesium phosphide for quarantine fumigations. Most uses of methyl bromide have been phased out but a quarantine fumigation is an allowed exemption for its continued use.

Methyl bromide is used for quarantine fumigation of logs, timber, and wooden materials (e.g., sawn timber, wooden packaging materials) that are notorious for carrying a variety of pests. These pests include insects that infest green and dry wood, nematodes, and some fungal pests of wood, notably oak wilt fungus (*Ceratocystis fagacearum*). Methyl bromide may be used to fumigate these commodities for quarantine/regulatory purposes by, or under the supervision of, a State or Federal agency. Phosphine, applied as a gas or as aluminum or magnesium phosphide, is an alternative to methyl bromide for treating logs, timber, and wooden materials, such as wood chips.

Chapter 7.2 Review Questions

Correct answers are given on page 175.

1. An example of an importation regulating agency that requires a quarantine fumigation to protect domestic agriculture and the environment from imported exotic pests is _____.
    a. the United States Environmental Protection Agency – Quarantine Department
    b. the United States Department of Agriculture – Animal and Plant Health Inspection Service
    c. the Occupational Safety and Health Administration – Human Health Program

2. If it will take more than 24 hours to fill a bin with grain for fumigation, you should fumigate _____.
    a. by continual addition of aluminum phosphide into the grain stream
    b. when the bin is filled to half its capacity
    c. once the bin is completely filled

3. Which of the following fumigants is used for the treatment of bulk stored grain?
    a. aluminum phosphide
    b. carbon dioxide
    c. metam sodium

CHAPTER 7.3

Burrow Fumigation for Vertebrates

> **LEARNING OBJECTIVES**
> - ☑ List any specialized equipment you need for burrow fumigations.
> - ☑ Briefly outline potential pests you can control with burrow fumigations.
> - ☑ Explain any unique label requirements for burrow fumigation.

> **Terms to Know**
> The following are important terms to know from this chapter. They are explained and *italicized* in the text and defined in the glossary at the end of this manual.
>
> Endangered Species
> Endangered Species Bulletins
> Threatened Species
>
> Toxicant
> Vertebrate

Burrow Fumigation

Fumigants may be used to control vertebrate pests in outdoor burrows. Vertebrates are animals with an internal backbone. You might use fumigants to control vertebrate pests such as Norway and roof rats, ground squirrels, moles, voles, and pocket gophers. The Vertebrate Pest Control Research Advisory Committee (VPCRAC) Online Vertebrate Pest Control Handbook, the University of California Statewide IPM Program's Pest Notes Library, or the University of Nebraska's Prevention and Control of Wildlife Damage, are just a few resources to help you correctly identify and manage these of vertebrate pests, including the suitability of burrow fumigation as a control method.

Devices that generate carbon monoxide (CO) to control burrowing rodent pests are available from several manufacturers. These devices are not fumigants, however many of the principles and practices of effective burrow fumigation below apply to the safe use of these devices.

General Principles for Fumigation of Burrows

Fumigants work best when soil moisture is high, such as in early spring, after soaking rains, or after irrigation. Moisture fills gaps between soil particles, which helps keep fumigant gases contained within the burrow system. Even when using best practices, burrow fumigation seldom achieves 100% control of the vertebrate pests in treated burrows. Follow-up treatments of active burrows may be necessary since fumigants have no residual effect and other vertebrate pests in the surrounding area can move into the treated burrows. Control as much ground as is occupied by the target species. Community based efforts to prevent re-occupation of treated burrows, instead of isolated treatments, improves long term success.

Registered Fumigants for Burrows

At the time of this manual's writing, there are three types of fumigants registered by U.S. EPA and DPR for use in outdoor burrows:
1. Ignitable gas cartridges,
2. Carbon dioxide: available in both solid pellets (dry ice) and as a gas, and
3. Aluminum and magnesium phosphide products.

Gas Cartridges
Some gas cartridge products are commercially available, but some products are only available to USDA-APHIS

Wildlife Services personnel. The gas cartridges come in the form of specially made cardboard tubes. These cartridges may contain a combination of sodium or potassium nitrate, carbon (usually in the form of charcoal), sulfur, and some other inert ingredients. The sodium nitrate accelerates the combustion of charcoal, which in turn forms carbon monoxide (CO)—a clear, odor-less, lethal gas poisonous to all animals that use hemoglobin in their blood to transport oxygen.

Carbon Dioxide
One product uses specialized equipment to pump pressurized CO_2 in a cylinder into burrows. Another type of product comes as a solid form of CO_2 ("dry ice") and is used specifically for rats and applied into their burrows. The solid CO_2 evaporates into a gas and suffocates the pest.

Phosphides
Some aluminum and magnesium phosphide products are labeled for use in burrows. These products are restricted-use pesticides and are California restricted materials, which require a permit from the County Agricultural Commissioner's (CAC's) office. Phosphine gas is highly toxic and capable of killing humans, vertebrate wildlife, and insects. Due to its toxicity, applicators must follow many guidelines to ensure safe and effective use. These include the use of air monitoring equipment, respirators, and developing a written detailed Fumigation Management Plan. These products also have label restrictions on how close they can be used to occupied structures.

Fumigation Practices
Many of the practices involved in preparing for a burrow fumigation are no different than those already discussed. Refer to Table 4-1: Fumigant Practices in Non-Soil Fumigation (Chapter 4, page 57) for practices applicable to burrow fumigations.

Advantages and Disadvantages of Fumigants for Burrowing Pests
Fumigants have some advantages over trapping or other *toxicants*, such as baits. They are easy to use—treat, then seal, active burrows. You do not need to perform pre-baiting or carcass search and removal. Fumigants also have several environmental advantages. Fumigants do not have any residual toxicity unlike many rodenticide baits. Fumigated burrows can be reoccupied after the fumigant has dissipated without risk of injury to the new occupant. However, this can also be a disadvantage since the pest problem may persist if other vertebrate pests return to and repopulate the burrows. Fumigants have a wide range of application temperatures allowing use throughout much of the year. Remember that the product label will provide information on the minimum temperature in which to apply the product.

Burrow fumigations have a few disadvantages. These applications can be costly. Researchers estimate that a burrow fumigation, compared to toxic baits, costs 5 to 10 times more in labor and/or product to treat the same amount of ground. Most burrow fumigations are used to control vertebrate pests on small acreages, sparse populations, or as a cleanup following the use of toxic bait. Another disadvantage of burrow fumigation is that the fumigant does not selectively target a specific vertebrate pest. Other animals in the burrow will be killed, whether they are targeted pests or not. As with other fumigants, burrow fumigations may pose health risks to both the applicators and bystanders.

Other Fumigation Considerations
Review the following chapters for key concepts applicable to burrow fumigations:
- Fumigant Management Plans: Chapter 4, "Fumigation Safety."

Conducting a Burrow Fumigation
Just like any type of fumigation, there are many factors to consider before the actual application of fumigant to a burrow. These include identifying the vertebrate pest, understanding its biology, surveying the site for both target and non-target animals, assessing the burrows for where is best to introduce the fumigant, and other considerations.

Surveying the Site
Burrow fumigants are highly toxic. Both target and non-target species may inhabit burrows and at times, may even occupy the same burrow. Label instructions contain guidelines designed to reduce the risk of killing non-target

animals especially *threatened* and *endangered species*.

A pre-fumigation premise inspection is an important step in the fumigation process and is usually required as part of the Fumigation Management Plan. For burrowing vertebrate pest fumigations, items to consider include:
- What vertebrate pests are present or nearby?
- What non-target species are present or nearby?
- Is the treatment area a potential habitat for endangered or threatened species? Is supervision by a person trained to distinguish dens of target and non-target species required?
- Where are the pests located?
- Is it the appropriate time of year to conduct a burrowing vertebrate fumigation (e.g., adequate soil moisture)?
- Are there weather conditions or other environmental factors which might impact the timing of the application? Generally, do not conduct burrowing vertebrate fumigations when it is too windy, too cold, or too dry.
- Is the soil type such that you will obtain a good seal (e.g., clay or loamy soil vs. sandy and rocky)?
- What equipment is needed for the fumigation?
- Are there factors outside of the target area to consider? (i.e., Are there occupied structures nearby? Could there be other employees or bystanders who may be near the fumigation site?)

The fumigant product you use may require you to survey the proposed treatment area for evidence of non-target animals. Surveys involve monitoring the site to determine if any non-target animals could be harmed by the treatment. The extent of these surveys will vary by product and application site. Some will simply require a visual inspection conducted in the morning and afternoon of the treatment area 24 hours prior to fumigation. Others may involve intensive multi-day surveys. A good visual inspection involves systematically and thoroughly viewing the area to identify all burrow openings, including the main entrance and all other exit holes. The inspection should also identify burrows/areas that should be avoided, such as those occupied by non-target animals including threatened or endangered species.

Assessing the Burrow

Even after surveying an area, you should always confirm that a burrow fumigation is indicated and appropriate for the vertebrate pest. Treatment of empty burrows, or burrows occupied by non-target animals, is not only illegal but also wastes time and money. You can determine if it is appropriate to conduct a burrow fumigation by evaluating several factors, including the presence of the vertebrate pest's scat (feces or droppings), the size and shape of the burrow entrance, and the burrow's architecture.

Many vertebrate pests are visible and can be easily identified during the day. If you are unable to visually confirm the presence of the target vertebrate pests when conducting your assessment, consider the diameter of the burrow openings you encounter. The average diameter of the burrow openings of some target species are:
- Ground squirrel: 4 to 6 inches.
- Pocket gopher: 2 ½ to 3 ½ inches.

Protecting Endangered Species

When the U.S. EPA determines that the use of a particular pesticide may harm a listed endangered species or its critical habitat without certain extra use limitations, it will spell out those extra restrictions in a bulletin.

U.S. EPA's application to access the bulletins is known as "Bulletins Live! Two." The labels of pesticides that fall under this system will direct you to download a required bulletin from the Bulletins Live! Two website.

The bulletins will list any special geographic restrictions on the use of the product.

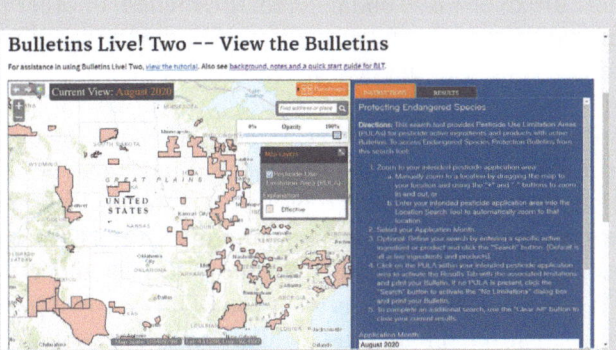

- Mole: 2 inches.

If you encounter a burrow with diameter openings that do not match that of the target vertebrate pest, do not proceed with the burrow fumigation until you have investigated and confirmed the presence of the target species.

Signs of Inactive Burrows
You can identify inactive burrows if they:
- Are completely or partially collapsed,
- Seem unkempt and unmaintained (e.g., spider webs or leaves and debris in entrance),
- Lack evidence of recent digging,
- Do not have fresh droppings/feces in and around the burrow.

Label Restrictions

All pesticide products labeled for burrowing pests can **ONLY** be used in outdoor burrows. Some labels will explicitly list the target pest species that the fumigant may be used to control. The use of the fumigant to control an unlisted pest species is a violation of the law (Figure 1).

The use of some carbon dioxide dry ice products may be prohibited within 10 feet of inhabited buildings or require that the building be unoccupied for 72 hours after application. Locations for use of phosphide based fumigants are even more strictly regulated. You cannot use phosphide-based fumigants in rodent burrows within 100 feet of any building on single or multi-family residential properties, nursing homes, schools (except athletic fields), day care facilities, and hospitals. The labels of these products will contain this, or a similar, statement:

DIRECTIONS FOR USE

USE RESTRICTIONS
Before applying product, read entire labeling. Only use for sites, pests, and application methods specified on this labeling. It is a violation of Federal Law to use this product in a manner inconsistent with its labeling.

This product may only be used to control woodchucks (*Marmota monax*), yellow-bellied marmots (*Marmota flaviventris*), ground squirrels (*Spermophilus spp.*), black-tailed prairie dogs (*Cynomys ludovicianus*), white-tailed prairie dogs (*Cynomys leucurus*), and Gunnison prairie dogs (*Cynomys gunnisoni*) in open fields, non-crop areas, rangelands, reforested areas, lawns and golf courses. This product may only be used underground in burrow systems. It may not be used to fumigate buildings or other man-made structures. Do not use in or under buildings or near flammable material, or when either the soil or the vegetation in the area to be treated are in extremely dry conditions. USE THIS PRODUCT ONLY IN BURROW SYSTEMS KNOWN TO BE IN ACTIVE USE BY THE TARGET SPECIES.

Figure 1: Target species on a burrow fumigant label. This label excerpt above, taken from a gas cartridge label, limits the use of the product in active burrow systems to control only the specific vertebrate pests listed. The use of the fumigant to control an unlisted pest species is a violation of the law.

> "For burrowing rodent applications: The use of this product is strictly prohibited within 100 feet of any building where humans and/or domestic animals do or may reside on single and multi-family residential properties and nursing homes, schools (except athletic fields), daycare facilities and hospitals."

Protecting Endangered Species

There are over 1,000 endangered, threatened, or other protected species in California. The use of registered burrow fumigants in active burrows may be limited to qualified individuals who have been trained to distinguish the dens and burrows of target pest species from those of non-target species (Figure 2).

Product labels may also have a statement requiring applicators to consult the U.S. EPA's *endangered species bulletins* for the area where the treatment will occur. These bulletins specify geographically specific pesticide use limitations that are intended to protect threatened and endangered species and their critical habitats. These bulletins are available using U.S. EPA's Bulletins Live! Two Application. If you are unable to access the bulletins, the labels may instead require you contact the U.S. Fish and Wildlife Service (U.S. FWS) prior to applying the fumigant. A U.S. FWS biologist will inform you of any known protected species in your treatment area. Be aware of any local county restricted material permit conditions restricting or prohibiting the use of a burrow fumigant.

In California, the Department of Pesticide Regulation (DPR) has an Endangered Species Program to protect federally and state listed species from pesticide uses.

DPR's Endangered Species Program activities include:
- Mapping sites occupied by federally and state threatened or endangered species;
- Evaluating the risks from pesticides to endangered species and their habitats;
- Classifying risks from pesticides registered in California;

> Use of this product within the occupied habitats of the organisms listed immediately below is limited to qualified individuals who have been trained to distinguish dens and burrows of target species from those of nontarget species.
>
> **Fresno kangaroo rat (*Dipodomys nitratoides exilis*)** in Fresno and Merced Counties, California;
>
> **Giant kangaroo rat (*Dipodomys ingens*)** in Fresno, Kern, Kings, Merced, San Luis Obispo, Tulare and Santa Barbara Counties, California;
>
> **Stephen's kangaroo rat (*Dipodomys stephensi*)** in Riverside, San Bernardino, and San Diego Counties, California;

Figure 2: Location specific applications of burrow fumigants. This label excerpt, taken from a gas cartridge label, restricts the use of the product to "qualified individuals" and to control only the target species listed in certain counties in California.

- Developing protection strategies to minimize risks from pesticides, as needed;
- Updating and maintaining the PRESCRIBE online database application; and
- Providing public outreach and applicator training on endangered species and their habitats.

One of the tools of the program is PRESCRIBE (Pesticide Regulation Endangered Species Custom Real-time Internet Bulletin Engine), an online database that helps pesticide users determine if there are any endangered species or species' habitat in the vicinity of their pesticide use site, and the use limitations that apply to the pesticide product(s) they intend to use.

In California, habitats of birds, mammals, reptiles, amphibians, fishes, invertebrates, and many plants are interspersed with agricultural areas. Of the federally listed species in California, the San Joaquin kit fox has the greatest overlap with agricultural areas, mostly in the San Joaquin Valley.

DPR coordinates its endangered species protection strategies with the U.S Fish & Wildlife Service, the National Marine Fisheries Service, the California Department of Fish and Wildlife, the California Department of Food and Agriculture, and the CACs.

Respiratory Protection

It is important to know when you are required to use a respirator when applying aluminum and magnesium phosphide products.
- If the concentration of phosphine is unknown or is greater than 15 ppm, handlers must wear a NIOSH-approved self-contained breathing apparatus (SCBA).
- If the concentration of phosphine is known to be between 0.3 and 15 ppm, handlers must wear a NIOSH-approved full face canister respirator – phosphine canister combination.
- If the concentration of phosphine is known to be less than 0.3 ppm, no respiratory protection is required.

An applicator is required to wear the respiratory protection listed on the label. Safety monitoring with gas detection devices will help you to determine the type of respirator, if any, to wear.

Soil Conditions and Fumigant Dispersion in the Application Zone

A fumigant must spread through the tunnels to any side tunnels, food storage areas, and nests within the burrow system to achieve effective control.

Burrowing pest control may be less effective in sandy and rocky soils. Soil moisture is also important. Always ensure that the soil is moist enough before application. In addition to assisting in the production of phosphine gas, moist soils will help keep the fumigant in the tunnels. Dry soils will not contain the fumigant since the cracks and spaces in the dry soil will allow the fumigant to escape. This results in a less effective fumigation and a more dangerous situation for the applicator and bystanders as the fumigant moves into the air. It is best to fumigate when soil moisture is high, such as in springtime, to ensure both the activation of the metal phosphide and a good soil seal.

You can use the soil ball test to measure soil moisture. To perform the soil ball test, you dig into the soil to a point where you are level with the burrow system, grab a sample of dirt, and ball it up. If the soil sample does not hold together then the soil is too dry, indicating that applying the fumigant is not recommended. Always follow label directions on locating and plugging all burrow entrances to minimize, to the extent possible, the amount of fumigant that escapes the burrow.

Application Timing

Temperature, wind, and soil moisture may affect the timing of the application. In addition to fumigating when soil moisture is high, the habits of the target pests may also influence the efficacy of a fumigation during certain times of the year. For example, planning a fumigation to control California ground squirrels when they are hibernating or estivating (the equivalent of hibernation during the hotter times of the year) may not be practical because these animal block the openings of their burrow with soil, making it difficult to locate the burrows that should be fumigated. Fumigation to control both California and Belding's ground squirrels is highly effective in the early months of the year, though this time may be shorter for the Belding's ground squirrels compared to the California ground squirrel.

Application Rate

Always follow label instructions to ensure that you are using the correct dosage. One aluminum phosphide product label lists the dosage rate as 2 to 4 tablets or 10 to 20 pellets per burrow opening. Since pellets are 20% smaller than tablets and release 20% less gas, more pellets are needed than tablets to obtain the same amount of phosphine. The label may also indicate that you use lower rates for smaller burrows and/or when soil moisture is high, and higher rates for larger burrow systems and/or when soil moisture is relatively low. You may also need to consider temperature, humidity, and wind when determining the appropriate dosage rate.

Applying more than the dosage rate indicated on the label increases the risk of human exposure due to excess phosphine gas release. Using dosages more than permitted by the label is not only unsafe, but also against the law. In contrast, applying less than 2 tablets or less than 10 pellets of this product is wasteful and can result in the ineffective control of the pest and/or the development of pest resistance. Always check the dosage indicated on the label, correctly measure the fumigant, and accurately report the amount of fumigant used.

Fumigation Posting Requirements

For burrowing pest control fumigations with aluminum phosphide, the label requires fields to be posted with signs which must contain:

- The signal words "DANGER/PELIGRO" and the skull and crossbones symbol;
- The words "DO NOT ENTER/NO ENTRE";
- The product name and EPA registration number;
- A 24 hour emergency response number; and
- The contact number of the certified applicator responsible for the application.

The signs must be no smaller than 9" by 11" and must stand at least 18" high from ground. Signs must be made of substantial material that can be expected to withstand adverse weather conditions and all information must be legible. Signs should remain posted for a minimum of 2 days after the final treatment and be removed by the certified applicator or grower.

California regulations require that the words "DANGER/PELIGRO" and "KEEP OUT/NO ENTRE" stated on the sign to be readable from a distance of 25 feet. The signs must be visible at all points of entry to the treated area, including each road, footpath, walkway, or aisle that enters the treated field, and each border with any worker housing within 100 feet of the treated field. If there are no usual points of entry, then the signs must be posted at each corner of the treated area. If the treated area is adjacent to an unfenced public right-of-way, the signs must be posted at each end of the treated area and at intervals of no more than 600 feet along the border of the treated area and the right-of-way. The signs must be posted before the application begins but shall not be posted unless the application is scheduled to begin within 24 hours.

Carbon Monoxide Pest Control Devices

Carbon monoxide (CO) pest control devices are defined as any method or instrument using CO to prevent, eliminate, destroy, or mitigate burrowing rodent pests. These CO pest control devices are regulated as "pest control" in

California however, they are not labeled as fumigants. Although these devices are not labeled as fumigants, they are commonly used to control vertebrate pests and you therefore should be aware of California regulations pertaining to these devices. California regulations require users of CO pest control devices to:

- Use devices with a U.S. EPA Establishment Number.
- Not use a device inside a structure inhabited by people or livestock.
- Based on the type of burrow, not use a device within certain distances (between 50 to 100 feet depending on target pest species) of structures inhabited by people or livestock.
- Not use a device on a burrow known or believed to be contain a non-target vertebrate animal or for purposes other than to control burrowing rodent pests.

Additionally, whenever an employee uses one of these devices to perform pest control for hire or for a local government agency, the employer must make certain the employee wears protective eyewear and must also maintain records of the application, covering the items detailed in the regulation, for two years.

Chapter 7.3 Review Questions

Correct answers are given on page 175.

1. What is a requirement for a phosphide based burrowing pest fumigation?
    a. cannot use if soil is too damp, such as in springtime
    b. cannot apply to burrows deeper than 3 feet
    c. cannot use within 100 feet of a building where humans may reside

2. Which of the following is a disadvantage of using fumigants to control a burrowing vertebrate pest?
    a. applications can be costly
    b. there is no residual toxicity
    c. pre-baiting is not needed

3. Which of the following is a sign of an inactive vertebrate pest burrow?
    a. the burrow is partially collapsed
    b. fresh droppings are near the burrow entrance
    c. there are areas of recent digging near the burrow

CHAPTER 7.4

Fumigation of Certain Structures

LEARNING OBJECTIVES
- ☑ Describe the uses of structural fumigation.
- ☑ List any specialized equipment you need for structural fumigations.
- ☑ Briefly outline potential pests you can control.
- ☑ Explain any unique label requirements for structural fumigation.

There are a wide variety of things that could be considered a "structure." California law defines "structural pest control" as the control of household pests (including but not limited to rodents, vermin, and insects) and wood-destroying pests and organisms or such other pests which may invade households or structures.

The intent of the fumigation is the primary determinant of whether licensing by DPR or the Structural Pest Control Board is required. The Structural Pest Control Board issues the license for the fumigation of businesses and residences to control household pests (such as cockroaches, termites, powder post beetles, bedbugs, and rodents). DPR's Non-Soil Fumigation category does not cover these fumigant uses and, relative to this chapter, instead focuses on fumigation of a food or feed commodity. Also note there are some items (such as dunnage or furniture) which can be fumigated under either license.

This chapter will focus on fumigation of other structures, such as warehouses and factories, that are fumigated to control storage pests and that fall under a DPR's Non-Soil Fumigation category.

Pests Controlled
The range of possible pests is large, but typically they would include food and grain destroying moths and beetles, and other animals, including cockroaches and rodents, that might infest areas where foodstuffs are present.

Purpose
These structures are fumigated to eradicate pests and prevent infestation of products as they are moved between producers, warehouses, and retail markets. The principles of structural fumigations are similar to those for fumigating commodities: use or create a confined air space where a fumigant can be released to eliminate pests infesting the area or commodity.

Fumigants Used
The primary fumigant is usually sulfuryl fluoride, though other fumigants may have some uses in certain situations.

Fumigation Practices
Many of the practices involved in preparing to fumigate structures for storage pests are no different than those already discussed. Refer to Table 4-1: Fumigant Practices in Non-Soil Fumigation (Chapter 4, page 57) for practices applicable to the fumigation of certain structures.

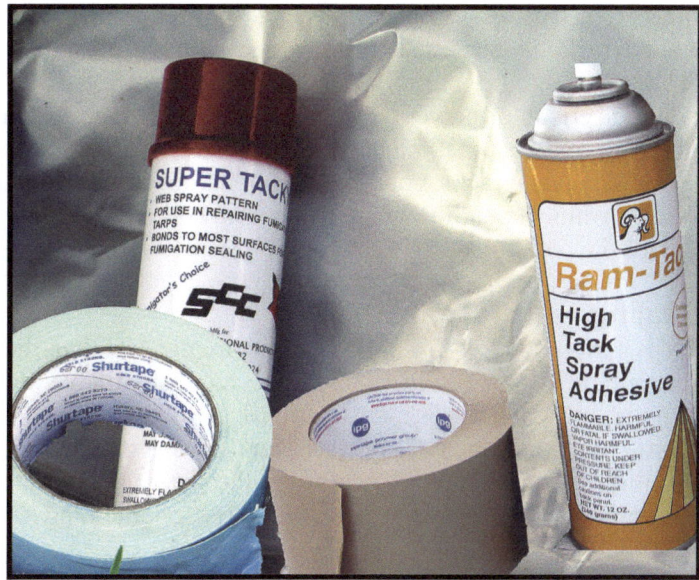

Figure 1: Tape-and-seal method of sealing. There are numerous sealing techniques and supplies that can be used for the tape-and-seal method but the most commonly used supplies include plastic sheeting, adhesive tape, adhesive sprays, and expandable foam caulking. (Photo © Betsy Danielson, Iowa State University Extension and Outreach).

Building Fumigation

Since most structures are not sufficiently airtight, they almost always require sealing. In relatively airtight structures, tape-and-seal methods might be sufficient to seal the building. However, many buildings will require you to use tarps over the entire structure. Any structure you fumigate needs to be sealed before introducing a fumigant. We will cover sealing in more detail in this chapter.

Sealing Buildings

Evacuate all occupants before you begin a fumigation. You might need to remove items that could be damaged by the fumigant. For example, the ProFume® label requires users to remove all non-target animals, desirable growing plants, drugs (including tobacco products), and medicines (including those items in refrigerators and freezers) prior to fumigation.

The importance of sealing cannot be overemphasized! Fumigant will leak from a poorly sealed structure, resulting in a failed fumigation and a health hazard to people outside the building. Be certain to close all doors and windows, seal chimneys and other openings, and repair cracks and broken windowpanes. Take special care to seal off adjacent storage or work areas that are not to be fumigated. Adjoining buildings sharing a common wall should be evacuated since it is difficult to guarantee that the fumigant will not diffuse to other areas.

Figure 2: Sealing with a tarp. It's very important to pad corners and other areas before tarping to protect tarps from tears. (Photo © Betsy Danielson, Iowa State University Extension and Outreach).

There are two methods to seal a building: tape-and-seal (Figure 1) and tarpaulins (tarps) (Figure 2). Tape-and-seal generally requires less sealing material than covering buildings with a tarp, It is also easier to aerate the building after the fumigation using this method.

Sealing with Plastic Sheeting

Plastic sheeting is most often used in a tape-and-seal fumigation. Plastic sheeting comes in a variety of thicknesses (measured in mils, a mil is one-thousandth of an inch, or 0.001 inch). Some are reusable, others are not. Thinner sheets (3 mil or less) should only be used once and for indoor treatments only. For outdoor treatments, you can use 4 and 6 mil polyethylene. 6 mil sheets can be reused if they are not excessively worn or ripped. However, it is easy for these sheets to get pin-prick size holes that are hard to see so be cautious when you reuse them. If you want to join sections of polyethylene, use fumigation tape instead of clamps, as clamps are more likely to cause tears. The tape-and-seal method can require a lot of time and effort to find and seal all possible leaks. It is very easy to overlook vents, cracks, conduits, and other openings where gas might escape. Also, fumigants can diffuse through outside walls depending upon their construction, making it difficult to maintain the gas concentration required to kill all the pests inside.

Sealing with Tarpaulins

Tarps generally refer to a more heavy duty material than the plastic sheeting used for tape-and-seal. Some tarps are specially made for fumigation. Commercially available tarps may be used if you have verified that they will contain the fumigant. Tarps made for fumigation are made of different materials including High Density Polyethylene plastic, vinyl, nylon, and polyester, and are designed to be used more than once. You can clamp sections of tarps together, so there is almost no limit to the size of the structure you can cover. When using more than one tarp, seal the seams by rolling the edges of two tarps together and securing the roll with clamps.

For any type of tarp, cover all sharp edges of the structure with material that can protect the tarp from tears (Figure 2). Mend or seal holes in the tarp with durable gas impervious tape.

Being well prepared is critical when fumigating an entire structure. Remove or protect all items that the fumigant may damage or that the label information requires removed or protected. Evacuate the building for the duration of the entire fumigation and the aeration period. Turn off pilot lights, flames, electrical heating elements, and electric motors.

Ground Seals

You need to make sure that tarps are sealed where they contact the ground. Often, people use snakes filled with sand or water to hold the edge of the tarp tight to the ground (Figure 3). Snakes prevent the tarp from blowing off during treatment. Double or triple the number of snakes if it is windy during treatment.

You can also use loose, wet sand or gravel for ground seals. Loose sand is especially important if there is danger of crushing or crimping gas sampling or introduction tubes where they enter underneath the tarp. If the soil where the tarp meets the ground is porous or dry, soak the soil around the perimeter of the structure with water. This will help prevent fumigant from escaping through the soil.

Prevent Billowing

Billowing of tarps can cause loss of fumigant from a structure. Billowing occurs when air beneath a tarp causes it to bulge outward. You can prevent this by keeping the tarp tight against the structure. For example, if the building top is flat, use sand snakes to hold down the tarp on the roof. If the roof is peaked, throw weighted ropes over the tarp. Draw the tarp as close to the building as possible. One technique to draw the tarp close to the building involves placing a high capacity electric fan in one doorway facing outward. This will create a partial vacuum that draws the tarp against the structure. You can then gather and tape down the excess material at the corners of the structure before removing the fan and sealing the doorway.

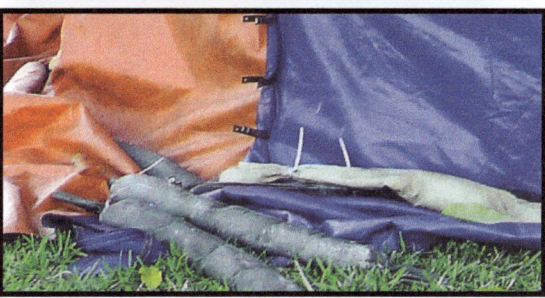

Figure 3: Ground seals: Snakes. Snakes are tubes of cloth or plastic filled about three-quarters full with sand, gravel, or water (top). When using snakes to hold the bottom of the tarp to the ground, they should overlap each other by about 1 1/2 feet. You should also use multiple snakes at corners (bottom). (Photos © Betsy Danielson, Iowa State University Extension and Outreach).

Other Structural Fumigation Considerations

Review the following chapters for key concepts applicable to fumigating structures:
- Choosing the proper fumigant: Chapter 1, "Fumigant Basics."
- Site characteristics that influence a fumigation, calculating fumigant dosages, air monitoring equipment: Chapter 3, "Planning for a Fumigation."
- Fumigant Management Plans: Chapter 4, "Fumigation Safety."
- Securing the site, warning signs, protecting applicators, and the public: Chapter 4, "Fumigation Safety," and Chapter 5, "Personal Protective Equipment."
- Application equipment and aerating the site: Chapter 6, "Fumigation Methods and Application Equipment."

Chapter 7.4 Review Questions

Correct answers are given on page 175.

1. When fumigating structures like warehouses and factories to control storage pests, the edges of two tarps can be joined and sealed with _____.
    a. duct tape
    b. foam
    c. clamps

2. If fumigation of a warehouse for storage pests shares a common wall with another structure, _____.
    a. the fumigation cannot be performed at any time
    b. the adjoining building must be evacuated before fumigation
    c. the fumigation should be conducted in the evening

3. Which of the following is **NOT** a method to seal a building for fumigation?
    a. tape-and-seal
    b. snake-and-seal
    c. tarpaulins

CHAPTER 7.5

Spot Fumigation

> **LEARNING OBJECTIVES**
>
> ☑ Describe the uses of spot fumigation.
> ☑ List any specialized equipment you need for conducting spot fumigations.
> ☑ Briefly outline potential pests you can control with spot fumigation.
> ☑ Explain any unique label requirements for spot fumigation.

Spot Fumigation

Spot fumigation is usually a short-term treatment of machinery or small storage spaces to control pests that infest whole foods and food particles that remain within processing equipment. Spot fumigation is used to control stored product pests in:
- Empty bins, silos, and holding tanks,
- Grain elevator boots heads, filters, conveyors, spouting, and purifiers,
- Food processing equipment such as sifters, rollers, dusters, and
- Related equipment in mills, food and feed processing plants, breweries, and similar industries.

Spot fumigation works by interrupting the life cycles of insect pests. Since one or more stages of the insect (egg, larvae, nymph, adult) may survive, you must regularly repeat spot fumigation to maintain control. Although possible, achieving total control of insect pests using spot treatment does not usually happen in actual practice. Factors contributing to less than 100% control are leaks, poor gas distribution, unfavorable exposure conditions, and insect harborages outside of the fumigation site. Also, some insects are less susceptible to certain fumigants than others. To achieve maximum control, you must take extreme care in sealing the application site. You also might need to use higher dosages as long as you do not exceed the maximum label rate, and/or extend the exposure period per label guidelines.

Advantages of Spot Fumigation
Insect infestations are usually not uniform. They concentrate in specific locations within equipment and storage areas. Spot fumigation allows you to treat only those areas where insects exist. This saves money and it puts less fumigant into the environment.

Disadvantages of Spot Fumigation
Spot fumigation is often labor intensive. You often must cut up and insert prepackaged fumigant into the machinery at several locations. After fumigation you must retrieve all the empty packages of fumigant or risk contaminating product during future processing. Spot fumigation can also be time consuming. Calculating the volumes of several small locations is cumbersome. Finally, disposal of spot fumigants, like incompletely reacted aluminum phosphide, is difficult because you must deactivate the fumigant before transporting it off site.

Pests Controlled
The target pests for spot fumigation are the same as those encountered in commodity fumigation since you are treating structures or equipment that hold or process food. These pests include food and grain destroying moths, beetles and other animals, including cockroaches and rodents, that might infest areas where foodstuffs are leftover.

Fumigants Used for Spot Fumigation
Some aluminum or magnesium phosphide, phosphine, and sulfuryl fluoride products are registered for use in spot treatments.

Fumigation Practices

Many of the practices involved in preparing for a spot fumigation are no different than those already discussed. Refer to Table 4-1: Fumigant Practices in Non-Soil Fumigation (Chapter 4, page 57) for practices applicable to spot fumigation.

Basic Spot Fumigation Procedures

Of the many factors that can affect the success of a spot fumigation, your understanding of both the equipment you will be treating and the airflow patterns within a structure are the most important. When developing an application plan, be sure to:

- Review the pesticide label prior to the application to identity applicable label requirements. Confirm that the intended use (application site) of the fumigant is indicated on the label, and that the planned dosage rate is within the prescribed rates. These will help to ensure an effective fumigation to control the target pest.

- Thoroughly examine the area to be fumigated and the equipment that will be used to determine if the treated area can be sealed sufficiently to hold gas. Review equipment schematics and/or diagrams when available.

- Assess the condition of the equipment for potential sources of leaks and determine the best sealing practice(s). You should also have a sound understanding of the facility design and the areas surrounding the site of fumigation.

- Establish and follow the Fumigation Management Plan to prevent exposure to any workers and bystanders during the fumigation and aeration.

- Establish a security plan to prevent entry by unauthorized and unprotected people into the fumigation area until after aeration. Post the required warning signs and notify the facility's personnel that a fumigation will occur.

- Ensure that any sensitive equipment, or their components that may contain metals that will react with the fumigant, are removed or protected from the fumigant. Phosphine gas can be corrosive to copper, copper alloys, and precious metals.

- Accurately calculate the volume within the space(s) to be treated. Since the volume of spot treatment area can vary widely, it is important you have a good estimate of the enclosed volume. If you will be using a tarp, first calculate the volume taken up by the equipment itself. After tarping is complete, revise the volume estimate based on the additional space contained within the tarp.

- Accurately calculate the dosage rates that will be used. Make sure your calculated rate is within the prescribed dosage rates specified on the product label.

- Make sure that the proper PPE, including respirators, are in good working order and on hand.

- Have proper monitoring equipment on site. Take measurements at regular intervals and note fumigant concentrations during the application to be sure that effective fumigant levels are reached and sustained long enough to kill the target pest. Have the proper monitoring equipment on site. Make sure that monitoring equipment has been calibrated and is in good working order. Measure fumigant concentrations during aeration to ensure gas levels are safe to allow reentry.

- If you are using aluminum or magnesium phosphide, have a plan for the recovery, deactivation, and disposal of the tablets or pellets.

Prior to the fumigation, clearly mark all application points. Also, mark points where ladders are needed to reach overhead areas. Prepare a chart of the structure that shows the location and number of application points. As you treat each point, check it off on your chart to ensure that you did not miss any points before moving on.

If equipment is present in the fumigated area, seal the equipment to prevent fumigant from escaping. Eliminate drafts inside the equipment by closing off sections that have openings and seal these openings with tape, caulk, tarps, or other appropriate materials. Failure to reach the effective concentration of the fumigant will result in an unsuccessful spot fumigation.

The safety measures when conducting a spot fumigation are equally important as with any other fumigation method. When possible, use a fan or hood to reduce your exposure. Read the label information of each product you use to determine the required PPE.

Use All Available Control Methods
Spot fumigation alone may not solve a pest infestation in some situations. Practice the principles of Integrated Pest Management. Sanitation is an important aspect in controlling stored product pests.

Other Fumigation Considerations
Review the following chapters for key concepts applicable to spot fumigations:
- Choosing the proper fumigant: Chapter 1, "Fumigant Basics."
- Site characteristics that influence a fumigation, calculating fumigant dosages, air monitoring equipment: Chapter 3, "Planning for a Fumigation."
- Fumigant Management Plans: Chapter 4, "Fumigation Safety."
- Securing the site, warning signs, protecting applicators, and the public: Chapter 4, "Fumigation Safety," and Chapter 5, "Personal Protective Equipment."
- Application equipment and aerating the site: Chapter 6, "Fumigation Methods and Application Equipment."

Chapter 7.5 Review Questions

Correct answers are given on page 175.

1. Which of the following is an advantage of spot fumigation?
    a. treatment can be applied to only areas where a pest is present
    b. broadcast application to large areas is possible
    c. total control of an insect pest is certain

2. Which of the following fumigants is registered for use in spot treatments?
    a. potassium phosphide
    b. carbon dioxide
    c. sulfuryl fluoride

CHAPTER 7.6

Fumigation of Transport Vehicles, Containers, and Ships

LEARNING OBJECTIVES

- ☑ Describe the uses of fumigation for transport vehicles, etc.
- ☑ List any specialized equipment you need to fumigate transport vehicles, etc.
- ☑ Briefly outline potential pests you can control by fumigating transport vehicles, etc.
- ☑ Explain any unique label requirements for vehicle and ship fumigation.

In this chapter we discuss some basic aspects of fumigating transport vehicles (i.e., boxcars, semi-trailers, shipping containers, and ships.)

The intent of the application is the primary determinant of whether licensing by DPR or the Structural Pest Control Board is required. As discussed earlier, in California "structural pest control" is the control of household pests (including but not limited to rodents, vermin and insects) and wood-destroying pests and organisms or such other pests which may invade households or structures. The Structural Pest Control Board issues the license for the fumigation of these structures. DPR's Non-Soil Fumigation category does not cover these fumigant uses and, relative to this chapter, instead focuses on fumigation of a commodity.

Applicators fumigating transport vehicles, which hold grains, lumber, or other commodities, to control storage pests need a DPR license or certificate with the Non-Soil Fumigation category. You need to be aware of some of the special regulations that come into play when fumigating these types of storage vehicles and structures. To name one, there are specific rules concerning whether you can or cannot fumigate certain transport vehicles while they are moving. These will be discussed in this chapter.

Pests Controlled
The target pests for the fumigation of transport vehicles, containers, and ships are the same as those encountered in commodity fumigation since you are treating storage structures that hold or held food. These pests include food and grain destroying moths, beetles, and other animals, including cockroaches and rodents, that might infest areas where foodstuffs are present.

Fumigants Used
The primary fumigants used for transport fumigation are aluminum or magnesium phosphide, phosphine, sulfuryl fluoride, and methyl bromide. Aluminum or magnesium phosphide, phosphine, and sulfuryl fluoride active ingredients are used to fumigate shipping containers, boxcars, and other transport vehicles. In the U.S., methyl bromide use is limited to quarantine and pre-shipment applications.

Fumigation Practices
The practices involved in preparing for a transport fumigation are no different than those already discussed. Refer to Table 4-1: Fumigant Practices in Non-Soil Fumigation (Chapter 4, page 57) for practices applicable to the fumigation of transport vehicles, containers, and ships.

Fumigation Environment Considerations
You already know that not all fumigants will give adequate control of the target pest in a grain fumigation, and that temperature and moisture can affect a grain fumigation. For example, temperature and moisture affects how quickly aluminum phosphide tablets or pellets are converted into phosphine gas. When moisture and temperature of the fumigated commodity are high, the fumigation could be completed in less than 3 days. However, at lower ambient temperatures and relative humidity levels, it might require 5 days or more.

Fumigation of Transport Structures

Fumigation of transport vehicles, containers, and ships prevents pests from being transported to other locations and protects shipped products from pest damage during transport. Items shipped in boxcars, truck trailers, shipping containers, ships, and barges are often fumigated after they are loaded into the vehicle.

Boxcar and Truck Fumigation

Fumigation of boxcars and truck trailers must comply with the regulations of state and local highway departments, and departments of transportation as well as fumigant label instructions. You can fumigate empty and loaded containers, trucks, boxcars, and other transport vehicles essentially the same way as other storage facilities—while they are stationary. As with any fumigation, you must ensure that the structure you will fumigate is airtight by sealing, taping, covering with tarps, or other appropriate methods. You must also follow all label, and state and federal requirements for fumigation such as the proper placement of warning signs, the use of proper locking techniques, completing required documentation, etc.

Ships, Barges, and Other Watercraft

Various watercraft carrying commodities are also fumigated. Sulfuryl fluoride and other fumigants are commonly used for these commodities and/or quarantine shipments. Fumigation of vessels of 15 gross tons or less and not engaged in passenger service must follow the same procedures used for fumigating structures on land.

Sealing

The watertight exteriors of boats allow for tape-and-seal fumigations. However, it may be more efficient to use tarpaulins to seal smaller boats or those that have a low profile. When fumigating cargo holds, seal all openings to the hold or tank and lock or otherwise secure all openings and entryways that might be used to enter the hold. Also, seal vents, piping, bilge openings, hatches, or other openings leading into the cargo hold. Place fumigation warning signs at all entrances to the treated spaces.

The fumigation of products or commodities on board a vessel depends on the type of infestation as well as the structure of the vessel. In most cases, a ship's cargo can be fumigated:
- In warehouse or storage silos before loading.
- In freight containers before loading.
- In the hold of a ship with fumigation and aeration complete before sailing.

Prior to fumigating a vessel for cargo fumigation, the master of the vessel (or their representative) and the fumigator must determine whether the vessel is suitably designed and configured to allow for safe occupancy by the ship's crew throughout the duration of the fumigation. If it is not safe, you cannot fumigate the vessel until all crew members are removed. Crew members cannot reoccupy the vessel until it has been properly aerated, and the fumigator and master of the vessel have determined it is safe for occupancy.

The certified applicator-in-charge of the fumigation must notify the master of the vessel of requirements relating to:
- Personal protection equipment,
- Emergency procedures,
- Cargo ventilation,
- Periodic monitoring and inspections,
- First aid measures,
- Detection equipment, and
- The requirement that a person qualified in the use of the fumigation equipment must accompany the vessel with cargo under fumigation.

> **Refrigerated Trailer Advantage**
>
> Refrigerated trailers are preferable to other trailers for fumigation as they are tighter and do not have a wood floor that a fumigant will leak through. In cool weather, thermostats can be turned up to achieve a more desirable temperature conditions for fumigation.

Ship Fumigation Regulations

Shipboard, in-transit ship, or ship hold fumigations are also governed by U.S. Coast Guard Regulations. Refer to 46 CFR 147A prior to conducting a fumigation. You must obtain special permits from the Coast Guard before you can fumigate barges.

Applicators who fumigate ships may need to go through a background check established by the Maritime Security Transportation Act. The Transportation Worker Identification Credential (TWIC) program requires a security threat assessment for people, including fumigators, who have access to secure areas of maritime facilities (ports) and vessels, followed by issuance of a biometric ID credential commonly called a TWIC card. Workers involved in the fumigation without TWIC credentials must always be escorted by someone who possesses these credentials.

In-Transit Fumigation

There are specific and stringent rules for fumigating in-transit (vehicles or containers moving from place to place):

- You can **never** use methyl bromide or sulfuryl fluoride in-transit (even while piggybacked on rail). Structures fumigated with methyl bromide or sulfuryl fluoride **must always** be stationary during the entire fumigation **and** aeration process.
- The labels for aluminum and magnesium phosphide, and phosphine gas, may allow fumigation of boxcars and containers, trucks, vans, and other transport vehicles shipped piggyback by rail in-transit. However, even when using these fumigants, it is illegal to fumigate and move vehicles over public roads or highways until they are fully aerated.

Warning Signs

As with any fumigation, applicators must make sure they affix warning signs to transit vehicles before fumigating transport vehicles. Warning signs must be made of substantial material that can be expected to withstand adverse weather conditions. Because these warning signs will be on moving vehicles, they need to be securely attached so they do not fall off during transit. For boxcar hopper cars, warning signs must be placed securely on both sides of the car near the ladders and next to or on the top hatch into which the fumigant is added.

Notification to Receiver

The certified applicator-in-charge must follow the specific requirements on the label of the product approved for in-transit fumigation. The shipper and/or certified applicator must provide written notification to the receiver of boxcars, shipping containers, and other vehicles that have been fumigated in transit. The certified applicator must provide a copy of the fumigation product's applicator's manual to the person receiving the shipment. The applicator's manual can either be sent ahead or included with the fumigated shipment. If the product applicator's manual is sent with the transport vehicle, the certified applicator must secure the manual on the outside of the vehicle. Proper handling of treated boxcars at their destination is the responsibility of the consignee. Upon receipt of the fumigated vehicle, a certified applicator and/or persons with documented authorized training must supervise the aeration process and the removal of the warning signs.

Shipboard

If the ship leaves port before the fumigation is completed or the cargo aerated, the person in charge of the vessel must ensure that proper PPE and gas or vapor detection devices, as well as a person qualified in their operation, are on board during the voyage.

Other Fumigation Considerations

Review the following chapters for key concepts applicable to transport vehicle, container, and ship fumigations.

- Choosing the proper fumigant: Chapter 1, "Fumigant Basics."
- Site characteristics that influence a fumigation, calculating fumigant dosages, air monitoring equipment: Chapter 3, "Planning for a Fumigation."
- Fumigant Management Plans: Chapter 4, "Fumigation Safety."
- Securing the site, warning signs, protecting applicators, and the public: Chapter 4, "Fumigation Safety," and Chapter 5, "Personal Protective Equipment."
- Application equipment and aerating the site: Chapter 6, "Fumigation Methods and Application Equipment."

Chapter 7.6 Review Questions

Correct answers are given on page 175.

1. If fumigating a vehicle in-transit, which of the following must **NEVER** be used?
    a. phosphine
    b. methyl bromide
    c. carbon dioxide

2. If it is not safe for the ship's crew to safely occupy the vessel throughout the duration of the fumigation, fumigation cannot be conducted until _____.
    a. the County Agricultural Commissioner approves a permit
    b. all crew members are removed
    c. a Fumigation Management Plan is developed

CHAPTER 7.7

Sewer Line Root Control

LEARNING OBJECTIVES

☑ Describe the problems roots can cause in sewer pipes.
☑ Explain the characteristics of roots and how they enter sewer pipes.
☑ Identify control strategies and formulations used for root control.
☑ Describe the PPE you need to wear when performing root control applications.
☑ Explain the basic application techniques and methods for root control in sewer pipes.
☑ Explain concerns with root control applications.
☑ Explain how to determine the effectiveness of root control management.

Terms to Know

The following are important terms to know from this chapter. They are explained and *italicized* in the text and defined in the glossary at the end of this manual.

Contact Herbicide	Slope	Storm Sewer
Systemic Herbicide	Grade	Combined Sewer
Selective Herbicide	Flow	Design Flow
Adhere	Sanitary Sewer	Actual Flow

Purpose of Sewer Line Root Control

One of the most destructive problems encountered in a sewer system is the growth of roots into sewer pipes. Roots can block or reduce flow, cause overflows, or reduce hydraulic capacity that leads to a loss of self-scouring flow. Sewer pipes with blocked or reduced flow often have septic pools that produce hydrogen sulfide (H_2S) and other dangerous or odor causing gases. Roots can damage pipes and other structural parts of a waste collection system.

Because sewer systems are underground and out of sight, most municipalities and homeowners only find out about sewer line root problems when stoppages and overflows occur (Figure 1).

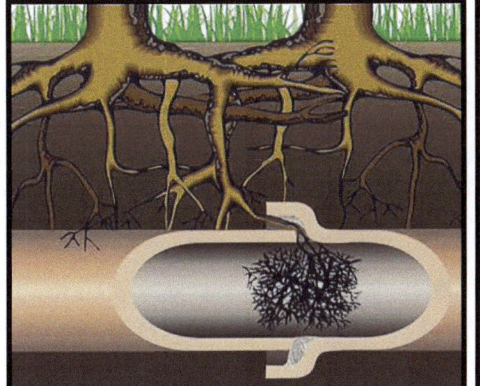

Figure 1: The importance of sewer line root control. Upon reaching a crack or a pipe joint, tree roots will penetrate the opening to reach nutrients and moisture inside the pipe (left). Left untreated, roots can completely fill and even burst pipes (right). (Illustrations from search.usa.gov. Public domain).

Although the term "root control" is most often used in relation to roots and sewer pipes, it is better to regard the concept as "root management." The concept of sewer line root management is to reduce the frequency and size of root intrusions into sewer pipes. This will reduce the frequency of sewer stoppages and overflows.

There are many effective means of managing roots in pipes including cultural, physical, mechanical, and chemical methods.

Pests Controlled
In pest management and control, accurately identifying the pest involved is the first and most important step. Pest identification in sewer line root control is simple since all roots in a sewer line are considered "pests." Unlike other fumigations where identifying the pest species (insect, plant, etc.) can factor into the treatment strategy, the tree species producing the nuisance roots is irrelevant in sewer line root control.

Be aware that sewer blockage may not always be due to root infiltration. Although determining the cause of a sewer blockage is difficult since the sewer is underground, make sure that fumigation to fix the problem is warranted.

Fumigants Used for Sewer Line Root Control
The primary fumigant used for sewer line root control is metam sodium. Metam sodium when mixed with water rapidly breaks down to MITC gas. MITC gas penetrates root masses to kill roots. Metam sodium for this use is often applied with dichlobenil, an effective growth inhibitor.

Fumigation Practices
Many of the practices involved in preparing for a spot fumigation are similar to those already discussed throughout the manual. Refer to Table 4-1: Fumigant Practices in Non-Soil Fumigation (Chapter 4, page 57) for practices applicable to sewer line root control.

Roots
Roots have three main functions. They:
- Anchor the plant and hold it upright,
- Store food for the plant, and
- Absorb and conduct water and nutrients.

Root Systems
Plants may have either a fibrous root system or a tap root system. Plants with fibrous root systems, such as garden plants and grasses, occupy the upper layers of soil and extend outward. These plants are not normally associated with sewer problems.

Trees and other woody plants usually have tap root systems. The primary root (called the tap root) of the plant grows directly downward into the soil. Tap root systems are well adapted to deep soils and soils where the water table is relatively low. Branches, or secondary roots, grow laterally from the primary root. Secondary roots can grow several inches in diameter and, if they invade sewer pipes, can exert enough pressure to spread pipe joints and break pipes.

Feeder roots are fine, hairlike roots that may develop into secondary roots. The surface of feeder roots have microscopic structures called root hairs. Root hairs increase the total surface area available to absorb nutrients and water. Upon reaching a sewer line crack or pipe joint, tree roots will penetrate the opening to reach the nutrients and moisture inside the pipe. This phenomenon continues in winter even though trees appear to be dormant.

Factors Affecting Root Growth
Soil conditions and seasonal changes affect root growth. When new pipes are constructed, the soil that has been excavated is backfilled around sewer lines. Backfill provides favorable conditions for root growth.

During the summer months, and with the lack of water, roots grow deeper to look for moisture. During the colder months, roots tend to search for warmer soils. Root growth is greatest during the fall, spring, and winter months, and less in the spring and summer months.

Root Growth into Sewer Pipes
In urban areas, tree roots may have a hard time finding good sources of nutrients. Limited green space, the removal of leaves and other organic debris from lawns, and the draining away of surface water by sewer systems cause roots

to seek water and nutrition at greater depths. Roots thrive in sewer pipes. Sewage systems are the perfect environment for root growth because they are well ventilated, oxygen rich, and full of water and nutrients. Microscopic openings in sewer systems can permit hair-like root structures to penetrate pipe joints, cracks, connections, or any other opening. A secondary root may grow alongside a sewer pipe for many feet, exploiting any opportunity to penetrate the system.

The roots of most trees cannot grow or survive if they are constantly submerged, so roots are usually not a problem in sewers that are located below a permanent water table. If the water table fluctuates, or if porous soil profiles allow rapid downward movement of rainwater, roots can be found in saturated soil and can be a major cause of sewer infiltration. In this case, tree roots suspended into the sewer system carry on metabolic activity while the woody, submerged, portion of the root systems serve as a pipeline for plant nutrients.

Types of Root Invasions
The two types of root invasions found in sewer lines are known as veil root structure and tail root structure. A veil root structure occurs in lines with steady, constant flows, such as interceptor pipes. The roots penetrate the pipe at the top or sides and hang from the upper surface like a curtain, touching the flow. The roots rake the flow and accumulate solids, debris, grease, and other organic materials. Eventually, the root mass and accumulated material can cause flow blockage, and gasses may be produced.

Tail root structures occur in sewers that have very low or intermittent flow, such as in small diameter collector sewers, building sewers, and storm drains. A tail root invasion looks like a horse's tail. The roots will grow into the pipe from the top, bottom, or sides, and continue to grow downstream, filling the pipe. Tail root structures measuring over 20 feet have been removed from sewers. Such root structures may appear as solid tubes of tree root, possibly with a slightly flattened area along the bottom where submergence in sewer flows prevents root growth.

General Concepts About Root Control in Sewers
Root masses are excellent collectors of grease and other solids. When these substances buildup, they can inhibit the effectiveness of root control pesticides by decreasing the contact of the pesticide with the root. There are many key factors and methods used in sewer line root control. Mechanical control is often used in conjunction with chemical controls to remove or prevent root growth.

You can determine root problems in sewer collection lines by:
- Reviewing maintenance histories that may indicate sewer lines that have experienced a stoppage and identified the cause of the stoppage.
- Reviewing sewer line video inspection reports that provide accurate evidence of a root problem.
- Identifying root prone areas. Sewer lines in the same area that were installed at the same time with similar nearby tree planting patterns may experience similar root problems.

There are conditions that increase the likelihood of root problems in a particular sewer section. These conditions include:
- Sewers located near other sewers with known root problems.
- Pipes located near the surface, closer to tree roots.
- Lines located off-road in wooded easements, in tree lined streets, or at the curb line near trees and roots. In general, sewer lines in residential areas are more susceptible to root problems than lines in industrial areas.
- Lines with many lateral connections per linear foot, affording greater opportunity for root intrusion.
- Sewer lines constructed with loose fitting joints or outdated joint packing material. Asbestos cement pipe, Orangeburg pipe, and clay tile sewers with oakum joints are very susceptible to root penetration. Pipe with air tight gaskets and seamless pipe are less susceptible.

A map (i.e., scattergram) of the local sewer system with known root problem lines highlighted is a useful tool for planning root control programs. As new problems arise and are highlighted on the map, patterns can emerge that indicate that an area is root prone.

Pesticides applied in the void area of the pipe with no surface to attach to, such as a root mass or pipe wall, enter the sewer flow and are transported directly to the wastewater treatment plant.

While chemical root treatments kill roots, they do not eliminate blockages. When a root mass is treated the roots may die quickly but can take weeks, months, or even years to decay and leave the system.

Methods to Control Roots

Root control methods can be grouped into non-chemical controls and chemical (pesticide) controls.

Non-Chemical Control Methods

Non-chemical methods, such as cultural, physical, and mechanical methods, are important tools for managing root growth. These methods, in conjunction with chemical (pesticide) control, can help mitigate and treat sewer line root issues as they arise to achieve the most effective outcome.

Cultural control methods can help mitigate future root problems. These control methods are routine management practices that can prevent roots from invading lines. This includes careful selection of the type of trees and plants to plant in an area and where to plant them, and ensuring that a thorough inspection of sewer pipes was conducted during construction. Municipalities should carefully inspect connections where plumbers join building laterals to main sewer lines. Also, homeowners should be advised of the potential for future root problems and should be discouraged from planting deep rooted or fast growing trees near sewer lines. **When a sewer root problem is detected, it is usually too late for cultural control.**

Physical control involves isolating the sewer pipe from the roots around or near the sewer pipe. Three examples of physical control are removal of plants and trees impacting pipes (if possible), replacing pipes, and pipe re-lining.
- Tree removal works best when the roots of a single troublesome tree have invaded pipes. It is often difficult to convince homeowners to agree to a tree's removal. Removal is risky because it does not guarantee the death of the roots. For tree removal to be effective, the stump should be pulled or treated with a basal bark application herbicide.
- Pipe replacement involves removing old, defective sewer lines and laying down new pipes. The new sewer lines should have airtight joints and properly installed connections to prevent roots from becoming a problem. Pipe replacement, however, is costly, often disrupts traffic, and damages property and trees near the sewer line. Also, roots can still enter through non-replaceable building sewers. If a pipe is in danger of collapsing or starting to fail, replacement may be the best control method.
- Sewer lines can also be rehabilitated by pipe re-lining. One method, "slip-lining," involves pulling a seamless pipe through the existing sewer and digging only where building laterals require connecting. Another method, "cured-in-place" lining, involves inflating and curing a sock or plastic tube that conforms to the shape of the pipe. Robotic devices are then used to cut building connections into the liner. Pipe re-lining can address root infiltration problems and correct some structural defects and is less disruptive than pipe replacement. It also can provide long-term control against root regrowth through joints. However, re-lining often is more costly than replacement, and roots can still enter though building laterals. Even after re-lining sewer main lines, pesticides may be required to prevent roots from entering through service connections.

Mechanical control is the most frequently used non-chemical control method and utilizes tools and machines that can discard and cut out roots from sewers and other pipes. It is the only method for relieving a root blockage since pesticides are ineffective and dangerous when used in plugged or surcharging sewers. Sewer stoppage is an emergency, and every municipality should have some type of mechanical control device for correcting the problem. The main disadvantage of mechanical control is that it provides no residual control or long term effectiveness. Root masses grow back heavier each time they are cut. Tap roots continue to grow in diameter and, in time, place stress on sewer pipes. Good results are obtained if the roots are cut flush with the joints, however, offset joints and cut in laterals can prevent the use of full gauge cleaning tools. Mechanical control is often used in conjunction with chemical control and physical control (e.g., preparing sewer lines for pipe re-lining).

Examples of mechanical controls are:
- Drill machines. These are hand or power driven flexible steel cables that turn augers or blades within the sewer. They are most often used by plumbers to relieve blockages in house lines or other small diameter sewers, and rarely in main line sewers.
- Hydrojetters. These use high pressure water and a special nozzle to first propel the nozzle forward to break up

the blockage. As the hose and nozzle are retrieved, the debris are flushed back, and the nozzle can be used to scrub the line.

- Rodding equipment. These work by connecting a special flexible cable or rod to rotating blade cutters, augers, or corkscrews and into a pipe or drain. They work by grinding up tree roots to clear the blockage. Rodding machines are most effective in small diameter sewers, up to 12".
- Winches. Large, engine driven winches pull buckets, brushes, or porcupine like scrapers through the sewer. Winches are most often used in large diameter sewers that cannot be cleaned efficiently with hydrojetters. Winches are useful for heavy cleaning to remove large volumes of solids.

Hydrojetters and rodding equipment can also be used to move the hose dispensing the metam sodium foam to the proper location in the pipe.

Chemical Control of Roots

Herbicides are used to chemically control roots. They can kill roots beyond the point of contact, providing control of root growth even outside sewer pipes. Herbicides kill plants or plant parts either by contact or systemic action. *Contact herbicides* act locally and cause quick dieback only of the parts of the plant that are exposed to the pesticide. *Systemic herbicides* are absorbed by roots or foliage and are carried throughout the plant. They take time, two weeks or more, to achieve desired results.

Herbicide activity is either selective or non-selective. *Selective herbicides* kill only certain types of plants and are used to reduce unwanted weeds without harming desirable plants. Some herbicides, for instance, affect only broadleaf plants, while others affect only grasses. Non-selective herbicides kill all plants contacted when applied at the label rate. They are used when no plant growth is wanted. Refer to The Safe and Effective Use of Pesticides, 3rd Edition, Chapter 10 (Whithaus, 2016), for additional information on the types of pesticides and their properties.

Many pesticide active ingredients are, or were, used to control roots in sewers. These include diquat dibromide, bensulide, dichlobenil, endothall, metam sodium, paraquat, trifluralin, 2, 4-D, 2, 4, 5-T, copper sulfate, and chlorthiamid. Be aware that not all of these pesticides are currently registered for use in California for sewer root control, nor are all of them fumigants. Before using any pesticide for root control, make sure that it is labeled for this use and registered in California. You may search DPR's Product/Label Database to determine the registration status of the product you intend to use.

This chapter focuses on the use of metam sodium for sewer line root control.

Metam Sodium

Metam sodium, the one fumigant used for root control, is a non-selective contact herbicide. Metam sodium root control products are formulated as an aqueous solution and are federal Restricted Use Pesticides. Metam sodium is a highly toxic pesticide as evidenced by the "DANGER" signal word on product labels.

Metam sodium is often used in combination with dichlobenil for sewer root control. After application, metam sodium breaks down into methyl isothiocyanate (MITC), a gas which kills plant roots. MITC is much more toxic than its precursor metam sodium. MITC may reach unsafe levels in poorly ventilated or confined spaces. Metam sodium is not systemic and does not move throughout the root system to kill the whole plant. Dichlobenil is used with metam sodium because dichlobenil is an effective growth inhibitor. Dichlobenil is formulated as a 20% or 85% wettable powder.

Metam Sodium Root Control Application Methods

In sewer line root control, applying metam sodium as a foam is the only approved method of using the fumigant. Applicators use specialized foam generating equipment to produce the foam, which is then applied to the interior of sewer pipes. Applications are made through hoses inserted to the length of pipe to be treated. While the hose is being retracted, foam is pumped in to fill the pipe with foam. As the foam collapses (over a period of 1 hour or more) it will tend to adhere to the pipe and root surfaces. We will discuss this more in the Application Techniques and Methods section of this chapter.

Foam may have the consistency of an aerosol shaving cream (dense, small dry bubbles) or that of dish washing soap suds (fluffy, large watery bubbles). For sewer use, the desired foam is usually that of an aerosol shaving cream. The drier phase of this foam is used to treat smaller pipe (less than 12 to 14 inches in diameter) and a wetter foam is used to treat larger pipe (over 14 inches in diameter). We will discuss foam quality in the Mixing and Calibration section of this chapter.

Foam is used to deliver root control pesticides because it:
- Effectively fills the pipe void above the flow line, contacting the pipe walls and root masses,
- Does not immediately break down after application, maintaining the required contact time for metam sodium,
- Prevents metam sodium vapor from drifting through the pipe into manholes and house vents,
- Contains surfactants and emulsifiers, which help herbicides penetrate through the grease and organic deposits on the root masses, increasing the effectiveness of metam sodium, and
- Allows treatment while pipes remain in service.

Once the roots have been killed (within hours of application), bacteria and other microbes in the sewer begin to break down the dead root tissue. Total decomposition of the roots may take several months to a year or more. The decomposed organic matter enters the wastewater stream and is carried to the water treatment plant. Root regrowth will start in a couple of years, which may make retreatment necessary at 3 to 5 year intervals.

Application Concerns
Metam Sodium Precautions

Due to the health risks involved with exposure to metam sodium products, special precautions must be taken to minimize exposure to bystanders (e.g., in a residence or business). The major concern is that the root pesticide foam will be inadvertently forced up service lines and into buildings, jeopardizing the health and safety of the people there. Labels include explicit directions to avoid backups. You can plug building drains to protect against backup and flooding. Accurate calculation of the application rate and careful application of the product will prevent the formation of excess foam that may then be forced up lateral lines and into building drains.

If metam sodium fumes have entered a building, evacuate the building. Metam sodium can be detected by its pungent, rotten egg odor. Open windows and ventilate with fans. Flush drains that emit the odor with ample water. If there is a spill or backup, the labels may direct you to use a squeegee, dustpan, or wet vacuum and garbage bags. Properly dispose of foam and liquid collected, wash the area with water and detergent, and flush the water down the drain. Absorbent materials (e.g., cloth, rugs, rags, etc.) that come into contact with the foam and liquid should be taken outside to dry before laundering separately.

Personal Protective Equipment for Metam Sodium Sewer Root Treatments

As always, follow the specific PPE requirements found on the label of the product you are using. We discussed PPE broadly for all fumigants in an earlier chapter but the PPE requirements for sewer line root control are unique. Below are some general PPE requirements for this application method.

Any worker who engages or carries out any operations likely to involve direct contact with metam sodium including mixing, loading, equipment calibration or adjustments, cleaning and repair of application equipment, entering into treated areas, sampling, cleanup of spills, rinsate disposal, or any other activities likely to result in direct contact with the product, should wear the following PPE:
- Splash resistant eye protection,
- Body covering including shirt and long pants or long-sleeved clothing. When a closed system is not used, mixers and loader must also wear a chemical-resistant apron or cloth coveralls, and
- Chemical-resistant gauntlet type gloves and boots.

Any worker within six feet of unshielded, pressurized hoses containing metam sodium solutions should also wear the PPE listed above.

If you are operating or monitoring application equipment, you must wear the following PPE at all times:
- Chemical-resistant footwear, and
- Body covering including shirt and long pants or long-sleeved clothing.

Workers operating or monitoring application equipment must have the following PPE immediately available to them:
- Either a half face respirator with organic vapor cartridge(s) plus splash resistant eye protection or a full-face respirator with organic vapor cartridge(s). Respirators must be worn in case of emergencies or leaks when the pungent, rotten egg or sulfur like odor of metam sodium is detected.
- Chemical-resistant gauntlet type gloves. These must be worn when a handler is likely to make direct contact with the product.

Wastewater Treatment Systems

Since the quality and amount of wastewater going through wastewater treatment systems is always changing, the efficiency of different treatment facilities will also vary. It is important to understand the major components in handling wastewater to minimize the risks associated with the use of metam sodium in these systems. Metam sodium's potential for affecting the treatment process is directly related to the concentration reaching the treatment plant and the efficiency of the treatment process in that plant. Due to these concerns, product labels include a special use restriction statement (Figure 2) to notify the wastewater treatment facility plant operators on measures that should be taken to protect the treatment plant and employees.

Systems for treating wastewater usually have three major components: collection networks, treatment facilities, and disposal processes. When applying metam sodium for sewer line root control, there are concerns at each one of these components.

Collection Networks

Wastewater, from sanitary and/or storm sewers, is collected and transported through a series of lines, pipes, and pumps of many sizes. Typically, a *sanitary sewer* line coming into the plant carries municipal wastes from households and commercial establishments and possibly some industrial wastes. Storm runoff is collected separately in a *storm sewer*, which normally discharges directly to a water course without treatment. In some areas, a single *combined sewer* picks up both sanitary and storm wastewater.

The collection network consists of a series of interconnecting pipes of varying sizes, from 4" diameter pipe to tunnels in which maintenance personnel can enter in boats. The pipe serving building is usually 8" - 12" in diameter. Collection networks are usually designed to use gravity to move the wastewater. Sanitary sewers are normally planned with a slope sufficient to produce a water velocity of approximately two feet per second or more when flowing full. Usually, this velocity will prevent the deposition of solids that may clog the pipe or cause odors. To allow maintenance personnel access to the collection network, it is broken up into sections by manholes. Manholes are usually spaced an average of 250 feet apart but can be 150 to 1,000 feet apart. Most treatment plants with flows of less than 0.5 million gallons per day have pipe sizes 4" – 8" and occasionally 10" – 12". As the plant capacities increase, the pipe sizes increase as lateral flows are collected and approach the treatment plant.

USE PRECAUTIONS NEAR TREATMENT PLANTS
Divide large scale applications to sewage collection systems in proximity to a treatment plant into smaller sectional treatments done at one or two day intervals to minimize effects on the sewage treatment process. Inform appropriate wastewater treatment plant officials prior to use so they may check for any unusual rotten egg or sulfur-like odor of metam-sodium above that of sewage and monitor the performance of filter beds or digesters. If the odor is detected at the sewage treatment plant or the biological breakdown process is adversely affected, cease root control applications until normal conditions are established. Reduce subsequent treatments in size or volume and spread over greater time intervals.

USER RESTRICTIONS NEAR TREATMENT PLANTS
This product must be used only where wastewater treated for root control will be processed through a wastewater treatment facility.

Applications must notify downstream waste water treatment facilities prior to the start of metam-sodium applications so that they may monitor the operations of the wastewater treatment plant.

Applicators must report how much product will be applied to the sewage system to operators of downstream water treatment plants and to inform these operators that high concentrations of these chemicals in wastewater may adversely affect the biological sewage breakdown process in wastewater treatment plants.

Never exceed the daily use of more than 15 gallons of PRODUCT for each million gallons of sewage flow (MGD) into the wastewater treatment plant (WWTP). Example: Inflow into the WWTP is 2.4 MGD, therefore, use a maximum of 36 gallons (2.4 x 15) of PRODUCT per day. When treating within one mile distance of the WWTP or when applying at night reduce the maximum application by 50% to 18 gallons (36 x .5). The above maximum daily use must extend over an eight hour work period.

Figure 2: User precautions and restrictions near treatment plants. Metam sodium labels for sewer root control require notification of wastewater treatment facilities. The label will also indicate restrictions related to the amount of product that can be used daily, the time of day the product is used, the distance of the application to the treatment plant, and more. As always, read and comply with the label directions.

Pipe slope, grade, and flow can all be variables that affect the velocity of the wastewater as well as the effectiveness of root control in wastewater collection systems.
- Pipe *slope* is an important design measure and is measured as the change in elevation between two manholes, and dividing that elevation change by the distance between the manholes. The slope of a pipe must be measured accurately to ensure the correct velocity rate when using pesticides.
- *Grade* is a measure of relative elevation from one area to another. For sewer lines, a building sewer is termed "below grade" if the elevations of floor drains are below the invert elevation of the upstream manhole. If a

building is "below grade," this can affect root control applications.

- *Flow* can affect patterns of root growth and is an important consideration in sewer line root control. How much wastewater flows through the system can be influenced by groundwater infiltration and can vary during peak periods of residential or industrial use. For sewer line root control, flow may dictate the appropriate treatment method, the rate of root decay following treatment, the rate at which pesticides move toward the treatment plant, and the rate of pesticide dilution in the wastewater stream. High flow can be detrimental if it dilutes the pesticide too much. It can also increase the movement rate of the root control pesticide toward the treatment plant, which can become problematic to treatment plants. However, flow can also help by diluting the pesticide before it reaches the treatment plant.

Foam should be injected above the flow surface to reduce the amount of pesticide carried downstream. Pipes with particularly heavy or swift flows should be treated at night or during periods of low flow to improve the efficacy of treatment. Be aware that if heavy or swift flows are problematic with respect to protecting the treatment plant, you must vary application rates accordingly. Also, be mindful of force mains upstream from the treatment area. Force mains above treated sections which "kick-in" after treatment can wash pesticides out of treated lines and move them downstream towards the treatment plant.

Large water users (such as industries) that contribute waste to the collection system may affect the efficiency of the treatment plant, especially if there are periods during the day or during the year when their waste flows are a major load on the plant. For example, some industries operate seasonally, making it possible to predict large flows. Even in areas where the sanitary and storm sewers are separate, infiltration of groundwater or storm water into sanitary sewers through breaks or open joints can cause high flow problems at the treatment plant.

The time required for wastes to reach a plant can also affect the efficiency of a treatment plant. Hydrogen sulfide gas may be released by anaerobic bacteria feeding on the wastes if the flow time is quite long and the weather is hot. This can cause odor problems, damage concrete in the plant, and make wastes more difficult to treat. Wastes from isolated subdivisions located far away from the main collection network often have these problems.

Pump stations are normally installed in sewer systems in low areas or where pipe is considerably deep under the ground surface. These pump stations lift the wastewater to a higher point from which it may again flow by gravity, or the wastewater may be pumped under pressure directly to the treatment plant. A large pump station located just ahead of the treatment plant can create problems by periodically sending large volumes of flow to the plant one minute and virtually nothing the next.

Dilution. Flow affects the rate of dilution of pesticides in the wastewater stream. The size of a wastewater treatment plant also determines the amount of dilution that occurs with wastewater at the plant. Concentrations of pesticides are measured in terms of parts per million (ppm) of active ingredient (A.I.) to water. One gallon of 100% A.I. mixed with 999,999 gallons of water represents one part per million.

Table 7.7-1. Parts per million equivalencies.

Parts Per Million (ppm)	Decimal Solution	Percentage
1 ppm	0.000001	0.0001%
10 ppm	0.00001	0.001%
100 ppm	0.0001	0.01%
1,000 ppm	0.001	0.1%
10,000 ppm	0.01	1.0%
100,000 ppm	0.1	10.0%
1,000,000 ppm	1.0	100.0%

The following example illustrates how to determine the amount of active ingredient, in parts per million, used in a sewer root control application.

The label instructions of a product that is converted into a foam for sewer root control indicate mixing 10 gallons of product (25% A.I.) in 200 gallons of water. This material is applied over the course of two hours to a sewer system with flows of 380,000 gallons per hour. To calculate the parts per million (ppm) of A.I. applied:

1. Calculate the gallons of product applied per hour.

   If 10 gallons of product is applied over two hours, then 5 gallons of product is applied per hour.

   $$\frac{10 \text{ gallons}}{2 \text{ hours}} = 5 \text{ gallons per hour}$$

2. Calculate the amount of A.I. in the 5 gallons applied per hour by multiplying the percent A.I. by the number of gallons applied per hour. Using the equivalencies in Table 7.7-1, the 25% A.I. in the product can be expressed as 250,000 ppm per gallon:

   250,000 ppm A.I. / gallon x 5 gallons / hour = 1,250,000 ppm A.I. applied / hour

3. Divide the amount of A.I. applied per hour by the flow rate:

   $$\frac{1,250,000 \text{ parts A.I. applied per million gallons (ppm) / hour}}{380,000 \text{ gallons per hour}} = 3.289$$

   3.29 parts of A.I. applied per million gallons (ppm) of water

The following example illustrates how to determine the amount of product that may be applied over time based on a target concentration.

An applicator using the same product in the previous example (25 % A.I.) learns from the wastewater treatment plant operator that the average daytime flow is 5,000,000 gallons, and that this is spread evenly over an 8 hour period in which the applicator intends to work.

If the concentration of the A.I. must not exceed 7 ppm so as not to present a health hazard to the waste treatment plant employees, what is the maximum amount of product can the applicator apply per hour to stay under 7 ppm?

1. First set up a ratio with 7 ppm equaling (x) gallons of A.I. per five million gallons.

   $$\frac{7 \text{ parts A.I.}}{1,000,000 \text{ gallons}} = \frac{(x) \text{ gallons A.I.}}{5,000,000 \text{ gallons}}$$

   Cross multiply and solve for x.

   7 x 5,000,000 / 1,000,000 = x

   x = 35

   The application can apply a maximum of 35 gallons of A.I. over the 8 hour period.

2. Next, determine the maximum amount of product the applicator can use over the 8 hour period. Calculate the gallons of product able to be applied by dividing the required gallons of A.I. allowed over the 8 hour period by the known A.I. concentration of the formulated product. Using the equivalencies in Table 7.7-1, the 25% A.I. in the product can be expressed as a decimal (0.25).

   $$\frac{35 \text{ gallons A.I.}}{0.25 \text{ A.I. per gallon}} = 140 \text{ gallons of product}$$

3. Divide the gallons of product by the amount of time the product is applied to determine the amount of product that may be applied over the specified period of time to stay under 7 ppm.

   $$\frac{140 \text{ gallons}}{8 \text{ hours}} = 17.5 \text{ gallons per hour of product with 25\% A.I.}$$

Treatment Plants

The downstream treatment plant size can be the most important factor in determining possible negative effects of the sewer root control treatment on a wastewater treatment facility. The size of the plant is measured in terms of the number of gallons of wastewater it treats per day. The daily flow for a community of 20,000 people (not including industrial and commercial uses) would be an estimated 1,600,000 gallons of water per day. Industrial and commercial use, as well as excess groundwater infiltration, could increase the number of gallons that a wastewater treatment facility treats per day. A treatment plant treating around 2 million gallons of water per day (MGD) is referred to as a 2 MGD plant.

It is important to distinguish between design flows and actual flows. *Design flow* is the amount of wastewater that the treatment plant is designed to handle daily. *Actual flow* is the actual volume of water that enters the treatment plant on a given day. If the actual flow of a treatment plant exceeds the design flow, the excess flows are bypassed around the plant and dumped directly into receiving waters, or stored temporarily in large basins.

Most flows in a sanitary sewer system occur during daytime hours. One half or more of the daily flow typically occurs between 6:30 – 8:30 a.m. and 4:00 – 9:00 p.m., although this estimate could vary depending on industrial use and groundwater infiltration. For low volume plants it is best not to estimate hourly flows based on typical usage. The treatment plant operator should be consulted for hourly flow rates.

Treatment Processes

Wastewater in a treatment plant flows through a series of processes to remove waste. The number of treatment processes and the degree of treatment usually depends on the uses of the receiving waters. Although not all treatment plants are alike, there are certain typical flow patterns that are common among treatment plants.

Wastewater entering a treatment plant usually flows through a series of pretreatment or preliminary treatment processes: screening, shredding, and grit removal. These processes remove the coarse material from the wastewater. Flow-measuring devices are usually installed after pretreatment processes to record the flow rates and volumes of wastewater treated by the plant. Pre-aeration is used to "freshen" wastewater and to remove oils and greases.

Generally, the next stage is for wastewater to go through primary treatment. During this stage, some of the solid matter carried by the wastewater settles out or floats to the surface where it can be separated from the wastewater. Secondary treatment commonly consists of biological processes. Organisms are used to partially stabilize (oxidize) organic matter not removed by previous treatment and convert it to a form easier to remove from the wastewater. Usually secondary treatment plants provide 3 to 30 hours retention time in the aeration portion of the treatment process. Retention time design is a function of plant size and plant type. A small extended aeration plant would probably require a 24 hour retention time. A 5 to 10 MGD plant might require 6 to 8 hours of retention. Waste treatment ponds, or lagoons, may be used to treat wastes remaining in wastewater after pretreatment, primary treatment, or secondary treatment. Before wastewater is discharged into receiving waters, it is usually disinfected with chlorine.

Disposal Processes

After completing treatment, the treated water is returned to the environment either through existing water sources or by using it for special types of irrigation. This water must meet health and contaminant level standards. Solid wastes must also meet contaminant standards before disposition. An excess of metam sodium in treated water or solid wastes may restrict treatment plant managers from disposing of these by-products and results in a disruption of normal operations.

Metam Sodium Effects on Treatment Plants

It is imperative to consider the size of a wastewater treatment plant that is downstream from an application of pesticides for root control. Metam sodium is a general biocide. It can adversely affect wastewater treatment plant facilities by disrupting the functional biological processes of the plant. Always consult the operators of a wastewater treatment plant located downstream from a sewer root control application site. Treatment plant personnel should also be made aware of the possible health effects of metam sodium exposure, such as the release of H_2S. Obtain as much information as possible about the treatment area from the treatment plant personnel. Important details include: the time of high flows, the size of sewer lines being treated, and the distance of the treated sewer line from the nearest lift station and sewage treatment plant.

A sewer line root control application must not adversely affect the daily operations of a treatment plant. Applications should not be made if there is a likelihood that the application can result in a problem at the treatment plant. Wastewater treatment plants depend on the growth and reproduction of microorganisms necessary to process wastes. Metam sodium can be toxic to these microorganisms. Treatment plant operators do not consider risking upset to their biological processes as an option and it must be avoided. Municipal treatment facilities can be fined up to $25,000 per day for National Pollutant Discharge Elimination System (NPDES) permit violations for "upset conditions." The NPDES permit program addresses water pollution by regulating point sources, including wastewater treatment plants, that discharge pollutants to waters of the United States.

Conduct a careful and thorough consultation with the wastewater treatment plant operator before applying any pesticide to a wastewater collection system. Maintenance and pre-treatment personnel at the plant must also be notified of the treatments. No two treatment plants are identical. Two plants with the same flow may react very differently to the same concentration of pesticides in wastewater flow if the biological process of one plant is under more stress (e.g., lack of oxygen, chemical pollutants, excessive organic loading, operator error) than the other. In a stressed treatment plant, a very small concentration of metam sodium could cause an adverse change in the biological decomposition process. This upset condition could last for days. Another plant without the additional stressors may be able to tolerate several ppm of metam sodium without any adverse effect. A well run plant is more tolerant of, and resilient to, pesticides. A manufacturer's recommendations for the use of their sewer root pesticide products in relation to wastewater treatment plants should only be used as a guideline. The best source of information about a given treatment plant and how it might respond to root control treatments is the treatment plant operator.

It may be necessary for the applicator to apply a reduced volume of the root control pesticide to prevent or minimize harm to the sewer system. Know the volume and hourly flows for the system and the manufacturer's recommended maximum concentrations so you can determine the maximum amount of product that can be injected into the system for any given day or hour. An application should be immediately discontinued if adverse effects (i.e., detection of the sulfurous odor, effects on the microorganisms essential to the plant's operations) are observed at the treatment plant. When applications are restarted, the applicator should use reduced application rates, specifically decreasing the total gallons of concentrate applied per hour (or day). The treatment plant operators should continue to monitor the plant to ensure that adverse effects do not reoccur.

Variables that may affect the impact of a metam sodium application on a specific treatment plant include:
- Type of pesticide and application method,
- Length, diameter, and slope of pipe,
- Distance and slope to treatment plant from application site,
- Size of treatment plant,
- Estimated flow (gallons per hour) at time of application,
- Existing problems at the treatment plant, such as those that compromise the efficiency of the plant's biological filtration process,
- The rate of breakdown of metam sodium in the wastewater.

Environmental and Endangered Species Considerations
Under certain conditions, metam sodium and MITC are highly soluble in water and can potentially leach and contaminate groundwater. Observe all precautions on the label to prevent potential environmental contamination.

Endangered and threatened species were discussed in an earlier chapter. Know that sewers may also harbor endangered or threatened species. Prior to a sewer root control pesticide application, use DPR's Pesticide Regulation's Endangered Species Custom Real-time Internet Bulletin Engine (PRESCRIBE) to determine if any endangered species are present in the area that will be treated.

Application Techniques and Methods for Sewer Root Control
As with any pesticide treatment, always use the correct application methods and accurately calibrated equipment. This will assure the most effective use of the pesticide, minimize the operational costs of the root control application, and protect the health and safety of the applicator, the environment, the public, and minimize negative impacts on the sewer collection system.

When assessing the various methods of application, the applicator must understand the principles of applying pesticides as well as the conditions that exist in the pipe being treated.

Application Equipment

The design and specific components of foam generating and application equipment can vary, but the operation principles are the same: the pesticide and wetting/foaming agent is diluted with water as per label instructions.

One type of equipment uses a mix tank (30 to 300 gallon) for diluting the pesticides with water. This mix tank is usually trailer mounted. Applicators then transport the mix tank to the various application sites throughout the municipality. One 200 gallon tank mix is sufficient to treat approximately 1,600 feet of 8 inch pipe. The pesticide/water solution is delivered under pressure (100 to 150 psi) to a foam production chamber. A positive displacement pump is then used to pump the pesticide/water solution.

A second type of equipment uses a smaller tank (3 to 6 gallon) to mix pesticides, but water is not added. This unit uses a pump to deliver water (without pesticide) under pressure (100 to 150 psi) to a venturi where the pesticide is introduced into the water stream and mixed just ahead of the foam production chamber. A venturi is a short piece of narrow tube between wider sections for measuring flow rate or exerting suction. The pesticide is then diluted with water during the application process. The pesticide/water dilution ratio is based on the product's concentration of active ingredient indicated on the product label. After the pesticide is mixed with water in the foam production chamber, air from a compressor at 80 to 140 psi is combined with the pesticide/water solution in the foam production chamber to produce the foam. The foam is then delivered to the interior of the pipe through hoses varying in size from 3/4" to 1¼"

Carefully review the product label to determine if there are equipment requirements specific to the product you will apply. For example, the label may require you to use a specific application nozzle designed to use water jetting to propel the application hose through the sewer. This allows water to escape under pressure and creates a rocket like effect that will both propel the nozzle forward, as well as prepare the roots and pipe for the application. Once the hose and nozzle reach the initial application point, turn off the water pressure and the spring loaded nozzle will switch to the foaming application. The applicator must retrieve the hose hydraulically while applying foaming herbicide within the pipe. The label may also require positioning a clean water rinse hose to continuously rinse off the application hose as it is extracted from the pipe, and to shut off the foaming herbicide 50 feet before the applicator nozzle reaches the manhole. This technique is used to purge the last of the foaming herbicide from the hose prior to it exiting the manhole.

Metam Sodium Application Techniques

There are several methods for making the application of foaming sewer root pesticides to sewers. Some of these techniques are briefly outlined below.

Hose Insertion Method

The hose insertion method is the most common way of applying foam to sewer lines (Figure 3). This technique has the lowest risk for unwanted foam traveling into laterals than other methods of foam application. A foam delivery hose is inserted through the section of pipe to be treated. Foam is pumped from a foam generator through the hose while it is being retracted at a predetermined rate (refer to the "Calibrating Hose Retrieval Rate" section of this chapter). Hydrojetters or rodding machines may be needed to move the hose into the pipe and position the hose before starting the foam application.

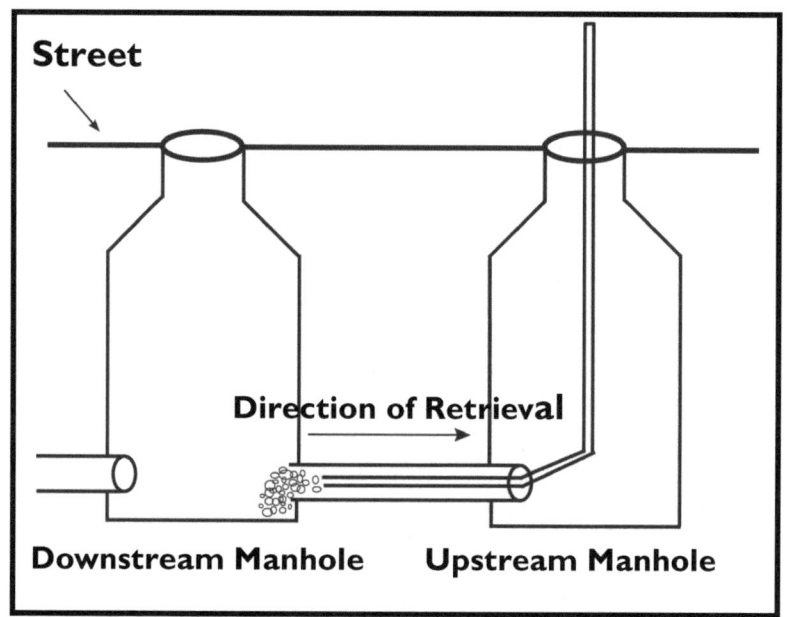

Figure 3: Hose insertion method. When using the hose insertion method of foam application, take care to avoid overfilling lateral lines. If buildings are close to the main, use cleanout plugs to prevent foam entry.

The insertion manhole may be upstream or downstream. However, whenever possible, the upstream manhole is the best for insertion, as this avoids drift toward the applicator. The pumping equipment is started once the hose reaches the other manhole. When foam begins to appear, the discharge hose is retracted at a specific rate. When using hydrojetters, it is recommended that moderate pressures be used when inserting the hose into the pipe. High pressures and excessive cleaning may result in excessive root damage, which can hinder the efficiency of the root control pesticides.

Hose Insertion Method: Split Treatments
Sometimes the sewer stretch may be longer than the amount of discharge hose, or it may not be possible to get the discharge hose completely through the sewer. In these cases, it may be necessary to use two set-ups to treat a section. With this technique, treat the downstream portion first, as this reduces drift toward the applicator.

Hose Insertion Method: Pushing a Slug
Foam will penetrate a distance beyond the discharge nozzle. If masses of roots or other obstructions do not permit the hose to be conveyed completely through the sewer, the equipment may be "allowed to pump" at a fixed location until the foam works its way through the obstruction. The equipment is then set up at the opposite manhole and the procedure is repeated until the two "slugs" of foam meet and overlap.

Hose Insertion Method: "Pulling the Water Out"
In some cases, sewer pipe may have inadequate slope, or swales in which water collects. As the foam is injected it displaces the water in the pipe. In these conditions it is often advisable to treat using the downstream manhole as the insertion manhole. As the hose is retrieved, excess water is pulled toward the insertion manhole. If the upstream manhole is used as the insertion manhole, then water may pond in the upstream manhole. If this happens, equipment must be shut down until water recedes.

Hose Insertion Method: Treating Service Lateral Connections from the Main
When treating a main line, often it is desirable to apply more foam to treat service connections (Figure 4). This provides an important benefit to property owners whose buildings are connected to these lines. Generally, treating service connections from the main is only feasible in small diameter (6" to 10") pipe. It is not possible to build up the pressures needed to penetrate service connections in large diameter pipe.

For this method, the applicator reduces the rate of hose retraction during the application process. Foam will always take the path of least resistance. With this method there is a risk of accidentally forcing the foam into buildings and you should use extreme caution to prevent foam from reaching building drains or outside sewer cleanouts. No more than 10 to 20 feet of service lateral should be treated using this method. Normal treatment of the main pipe is sufficient to kill roots at the service connection.

Factors that increase the risk of foam entering buildings:
- Basements with below grade plumbing,
- Floor drains,
- Dry traps,
- Reduced sewer volumes due to flow, low spots, or root masses,
- Unknown connections to the service lateral being treated.

Surface Coating Large Diameter Pipe
It is often impossible or too costly to completely fill the sewer with foam when treating large diameter pipe. Only the pesticide that contacts

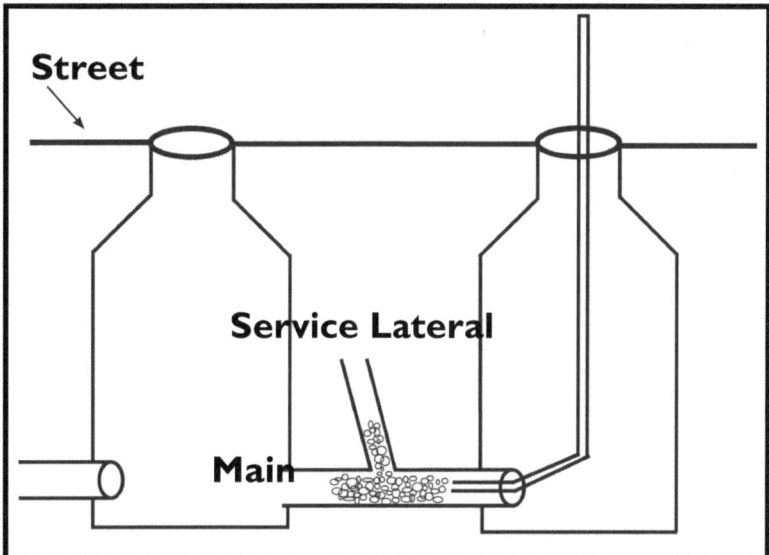

Figure 4: Treating service lateral or "Y" connections from the main. You will need additional foam per foot when using this method. Calculate the amount of additional foam you need for the given number, and pipe size, of building laterals, and vary retrieval rates accordingly.

roots is useful. To coat the interior sewer line surface, an elevated nozzle is pulled through the sewer. Foam is ejected through the nozzle above the flow where it contacts and sticks to pipe surfaces and roots. (Figure 5) It is very important that the nozzle be elevated above the flow. If the foam is ejected into the flow, it will not contact pipe surfaces. To calculate the volume of foam required to coat pipe surfaces refer to label instructions and refer to the example provided in the "Calculating Pipe Volume" section of this chapter.

Surface coating often does not yield the same results obtained by filling the pipe, since the foam is not under pressure and will not penetrate root masses as effectively as it would when filling the pipe. Repeat treatments may be necessary as succeeding layers of root tissue are killed. Also, surface coating will not penetrate service connections. Surface coating is also sometimes used on small diameter pipe when heavy flow rates preclude filling the pipe with foam.

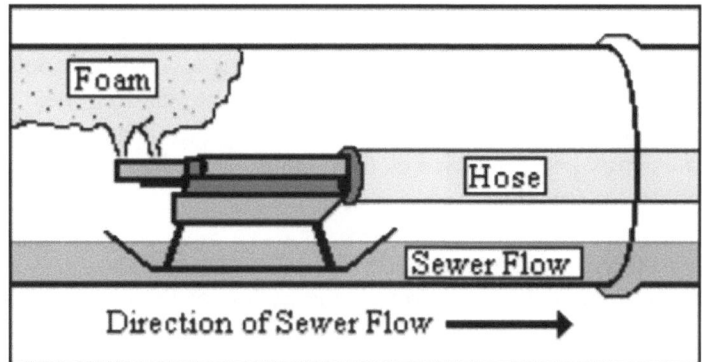

Figure 5. Surface coating large diameter pipe. Foam is ejected through the nozzle above the flow where it contacts and sticks to pipe surfaces and roots.

Spot Treatments

Spot treatments may be used either with foam filling or surface coating techniques. Spot treatments involve treating only places where the roots are. The advantage of spot treatments is that less pesticide is required to treat a given length of sewer pipe. The disadvantage is that it is first necessary to know exactly where the roots are, which requires a recent video inspection. If the video inspection was not performed recently, additional root penetration may have occurred that would miss treatment. Additionally, initial root penetrations frequently are unnoticed or missed by video inspection. Spot treatments are most useful in large diameter lines where the increased cost of material offsets the cost of video inspection. The amount of pesticide that can be saved on small diameter pipe is usually negligible and not worth the cost of video inspection. When using spot treatment techniques, allow a certain amount of overlap on each side of the root masses, approximately 10 feet to each side of the root intrusion.

Pushing Foam Through Inflatable Plugs

In some cases, it may be desirable to push the foam through inflatable, hose-through plugs (Figure 6). These plugs are available through plug vendors in the sewer industry. Insert the plug at one end of the line with the hose running through it. Inflate the plug and inject foam in a volume required to fill the pipe or until foam appears at the opposite manhole. When using this method, because foam always follows the path of least resistance, there is a significant hazard of foam backing into buildings. This method should only be used where there are no service connections on the main-line sewer or where buildings are set far back from the main.

Treating Building Sewer Lines

Extreme caution should be taken when treating building service lines. These pipes are connected directly to building drains and there is a chance of accidentally forcing the foam through the pipe and into the building. Service lines usually connect the building to a sewer main located in the street in front of the building or in an alley behind the building. These service lines are usually a 4" to 6" diameter pipe and were installed as buildings were erected. Records of where these lines go, which buildings are connected to a specific main, and pipe conditions, are usually non-existent.

Treating building service lines is usually done in one of two ways: treating the service line from the main as discussed earlier, or using a hose-through inflatable plug to treat from the building toward the sewer main.

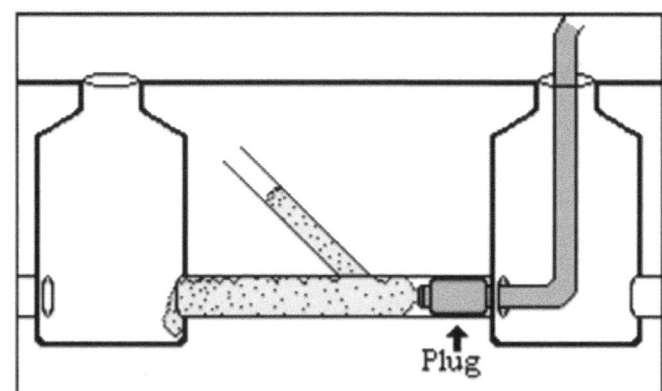

Figure 6: Pushing foam through inflatable plugs. Use this method with caution as it may cause foam to back up into buildings.

The inflatable plug method of service line treatment is generally less risky than treating from main lines. In principle, a specially designed plug is inserted in the service lateral between the point of treatment and the building. The plug is a 1" hose with an air bladder molded around the outside. The bladder is then inflated to block the foam from exiting the clean out or from being forced back toward the building. If there are cleanouts or fixtures downstream from the insertion cleanout, they must be plugged. Calculate the volume of foam necessary to treat the given distance of building sewer. The equipment pumps the foam through the hose and down the service lateral to the main until the desired quantity of foam has been pumped. Treat house lines only if you are familiar with the design of the building's sewer systems or are working under the supervision of a licensed plumber. Be cautious, improper application may result in foam being discharged into houses. Building occupants must be alerted that treatment is taking place and should be advised to exit the building if they detect the rotten egg or sulfur like odor of metam sodium.

Factors that increase the risk of foam discharge into houses when treating service laterals:
- Unknown connections to the service lateral being treated,
- Inserting hose into upstream, instead of downstream, pipe,
- When treating from within buildings, the plumbing may not have pipe branch connections between the hose plug (treatment location) and the main line.

Treating service laterals can be a high exposure risk procedure. If the exact configuration of service laterals is not known, do not treat. Do not treat service laterals from buildings which are inaccessible at the time of treatment. Spotters should be in all buildings when service lines are being treated to monitor for signs that foam is being forced into the building.

Mixing and Calibration

Due to differences in products, specific mixing instructions must be obtained from the label of the metam sodium root control product being applied. Mixing instructions must also be obtained from the equipment manufacturer for the specific equipment being used.

Mixing the Pesticides

Applicators may make the mistake of filling mix tanks from fire hydrants, garden hoses, or other fresh water sources. If there is a pressure drop in the water distribution system, the solution in the mix tank could back siphon into the fresh water system, contaminating the entire water distribution system. Whenever a tank is being filled with water it should never be left unattended. Back siphoning can be prevented with an air gap, a back flow prevention device, or an intermediate water source. Refer to The Safe and Effective Use of Pesticides, 3rd Edition, Chapters 7 and 10 (Whithaus, 2016) on how to safely mix pesticides, including using back flow prevention devices and air gap separations.

For sewer line root control, metam sodium and dichlobenil can only be used in combination with each other, as per product label instructions. As discussed in the section on application equipment, pesticides may be mixed with the water in a mixing tank, or may be mixed in a smaller tank and automatically mixed with water at the moment of foaming. Dichlobenil should be mixed vigorously with metam sodium before adding water. The mixed solution should not be allowed to stand. Mild agitation is necessary to keep the dichlobenil in suspension.

Be careful when mixing metam sodium with water. Remember that metam sodium quickly breaks down to the more volatile and toxic MITC when mixed with water. The solution should be used promptly after mixing. Do not mix more solution than will be used in one day. To determine the amount of solution needed:

1. Determine the method of treatment.
2. Determine the total footage of pipe and diameters of all lines to be treated.
3. Calculate the pesticide mix ratio and the amount of pesticide/water solution to prepare.

Calibrating Foam/Solution Expansion Ratio

Knowing how to properly calibrate application equipment is important to achieve the correct consistency and volume. Below are recommendations for equipment and foam calibration. The equipment manufacturer should be consulted for more specific details.

Calibration is important in a successful root control treatment. For foam fill applications, the industry standard expansion ratio is 1:20, indicating that 1 gallon of the pesticide/water solution will expand to fill 20 gallons of volume. You will know that the foam calibration is correct when the foam is dense and has small bubbles. Moreover, the foam will retain its shape and adhere to pipe surfaces. If a foam is less than the correct ratio, a thin "wetter" foam that is not as dense is produced and will not stick to pipe surfaces. A "dryer" foam with an expansion ratio greater than 1:20 has large bubbles and will not perform as well.

 NOTE: Some pesticide manufacturers have formulations intended to produce a wetter foam used for surface treating large diameter pipe. Instructions for these formulations generally specify an expansion ratio of 1:14 (14 gallons of foam from 1 gallon of pesticide/water solution).

Foam quality can be adjusted by modifying the pesticide/water solution. Follow the manufacturers guidelines to make foam quality changes. A foam has the correct foam quality if, once applied, it breaks into light balls and flakes of foam about 2 to 3 feet from the point of discharge. Tests for foam quality can be done prior to application at the treatment site by adding the correct amount of wetting/foaming materials without the pesticides. This reduces the risk of pesticide exposure for the operators performing the test. To confirm that foam has the industry standard expansion ratio of 1:20, fill a 2,000 milliliter (ml) graduated cylinder to the top with the foam and place it in an area that is not exposed to wind. After the foam has settled, for a 1:20 ratio solution, the remaining liquid should measure 100 ml.

Calculating Metam Sodium Solutions

To achieve effective sewer root management and at the same time protect people and the surrounding environment, you should know how to correctly calculate the amount and concentration of pesticide to use for the treatment of a sewer line. Calculating the amount of pesticide required varies depending on pipe size, dilution, length of pipe, and other variables. Tables 7.7-2 and 7.7-3 are worksheets for calculating the amount of pesticides needed using the foam fill method and the foam spray application method, respectively. To use these tables, enter the pipe length and the required dilution ratio specific to your application and perform the calculations from left to right. Both tables are for illustrative purposes only. The product label and labeling should be consulted to determine manufacturer's recommendations for actual amount of product required and for specific mixing and application instructions.

Use Table 7.7-2 to calculate the amount of pesticide needed for a foam fill application.

Table 7.7-2: Foam Fill Application Example

Pipe Diameter	Gallons Per/Foot	Length of Pipe	Gallons of Foam	Service Laterals 15-25%	Total Gallons Foam Required	Expansion Ratio 1:20 Required	Chem/Water Solution	Dilution Ratio Required	Total Product to Use (Round Up)
4"	0.7	x	=	÷	= ____	÷ 20	= ____	÷	____
6"	1.5	x	=	÷	= ____	÷ 20	= ____	÷	____
8"	2.5	x	=	÷	= ____	÷ 20	= ____	÷	____
10"	4.0	x	=	÷	= ____	÷ 20	= ____	÷	____
12"	6.0	x	=	÷	= ____	÷ 20	= ____	÷	____

For example, if you know that you are treating a 6 inch pipe that is 1,000 feet and the dilution ratio is 26 parts water to 1 part pesticide and the service lateral is 20%, you can calculate the **total amount of pesticide needed**.

Pipe Diameter	Gallons Per Foot	Length of Pipe	Gallons of Foam	Service Laterals 15-25 %	Total Gallons Foam Required	Expansion Ratio 1:20 Required	Chem/Water Solution	Dilution Ratio Required	Total Product to use (Rounded up)
6"	1.5	1000	1500 (1.5 x 1000)	300 (1500 x 20%)	1800 (1500 + 300)	÷ 20	90 (1800 ÷ 20)	÷ 26	3.5 (rounded up to 4.0)

The total product needed (or to be used), after rounding, is four gallons.

Use Table 7.7-3 to find the amount of pesticide needed for foam spray application.

Table 7.7-3: Foam Spray Application Example

Pipe Diameter	Gallons Per/Foot	Length of Pipe	Gallons of Foam	Expansion Ratio 1:15 Required	Chem/Water Solution	Dilution Ratio Required	Total Product to Use (Round Up)	Water to Use (gallons)
12-14"	3.0	x ____	= ____	÷ 15	= ____	÷ ____	= ____	
15-16"	3.5	x ____	= ____	÷ 15	= ____	÷ ____	= ____	
18"	4.3	x ____	= ____	÷ 15	= ____	÷ ____	= ____	
20"	4.5	x ____	= ____	÷ 15	= ____	÷ ____	= ____	
21"	4.75	x ____	= ____	÷ 15	= ____	÷ ____	= ____	
22"	5.0	x ____	= ____	÷ 15	= ____	÷ ____	= ____	
24"	5.5	x ____	= ____	÷ 15	= ____	÷ ____	= ____	
26"	6.0	x ____	= ____	÷ 15	= ____	÷ ____	= ____	
27"	6.75	x ____	= ____	÷ 15	= ____	÷ ____	= ____	

For example, if you know that you are treating a 12 to 14" diameter pipe that is 500 feet and the dilution ratio is 26 parts water to 1 part pesticide, you calculate the total amount of water needed.

Pipe Diameter	Gallons per foot	Length of Pipe	Gallons of Foam	Expansion Ratio 1:15 Required	Chem/Water Solution	Dilution Ratio Required	Total Product to use (Rounded up)	Gallons of Water to Use
12-14"	3.0	500	1500 (3 x 5000)	÷ 15	100 (1500 ÷ 15)	÷ 26	3.8 (rounded up to 4.0)	96 (100 − 4)

The required amount of water needed can be obtained by subtracting the total product required (4 gallons (rounded up from 3.8)) from the total amount of chemical/water solution calculated (100). For our example, the required amount of water needed is 96 gallons.

Calculating Pipe Volume

Knowing the correct pipe volume ensures that you use the correct amount of foam. If the pipe volume is overestimated, too much foam can be applied which can be wasteful, and if it is underestimated, the application may be ineffective.

Pipe volume is calculated by multiplying the radius (half the diameter of the pipe) squared by the length of the pipe and the constant π (3.14) (Figure 7).

Wastewater in a partially filled pipe takes up volume as well. The applicator should be able to compensate for the reduced volume. Roots do not grow below the water level and chemicals are not effective when diluted in sewer flows. Proper application requires that the foam be discharged above the flow line. Under certain conditions where a pipe has slow moving flow, the applicator should compensate for the volume displaced by the water in order to avoid overcharging the pipe.

Figure 8 illustrates how to calculate the portion of the pipe that is submerged. In the illustration, "D" corresponds to the diameter of the pipe and "d" corresponds to the depth of flow. The applicator needs

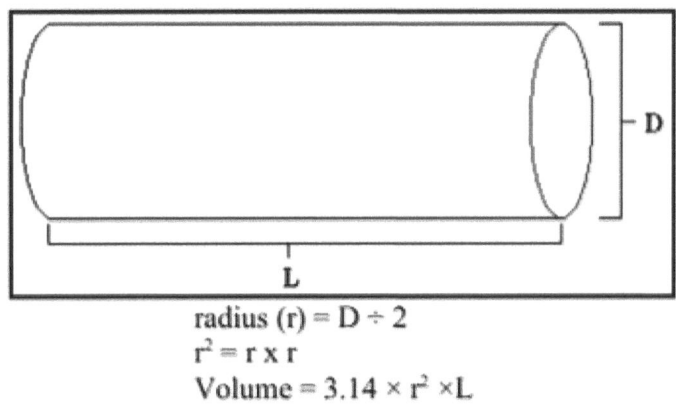

radius (r) = D ÷ 2
$r^2 = r \times r$
Volume = 3.14 × r^2 × L

Figure 7: Calculating pipe volume.

to know the relationship between d and D to estimate the percentage of the pipe circumference that is submerged. To do this, the depth of flow, d, is divided by the diameter of the pipe, D. The table on the right shows the amount of pipe circumference that is submerged (the "wetted perimeter") based on the calculated ratio. When calculating the amount of pesticide needed, you need to subtract the submerged portion of the pipe from the total (Pipe circumference to use for application calculation = Total pipe circumference – Submerged circumference).

Below is an example of determining the amount of foam needed for a surface spray application where there is some flow in the pipe.

If you are going to surface coat a 24" diameter pipe with a 7" deep wastewater flow with 3" of foam for 300 feet, how much foam will you need?

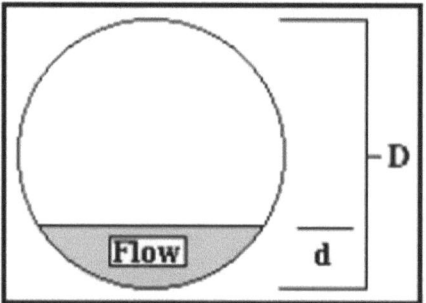

d/D	% pipe volume submerged
0.1	20%
0.2	30%
0.3	37%
0.4	44%
0.5	50%
0.6	56%
0.7	63%
0.8	70%
0.9	80%
1.0	100%

Figure 8: Calculating pipe portion submerged.

A. Calculate the circumference of the pipe. The formula for the circumference of a circle is 3.14 (π) x D. The diameter, D, of a 24" pipe, in feet, is 2'. Therefore, the circumference of the pipe in feet is:

3.14 x 2' = 6.28 ft

B. Estimate the % pipe volume submerged. The portion of the pipe submerged is d ÷ D where d=wastewater flow. Use the table in Figure 7 to approximate the percentage of the pipe that is submerged.

7" / 24" = 0.29

Using the table in Figure 7, 0.29, rounded to 0.3 is approximately 37% of the pipe that is submerged.

C. Determine the portion of the pipe circumference that is submerged by multiplying the circumference of the pipe (A) by (B) the % pipe volume submerged (in decimal form).

6.28 ft x 0.37 = 2.32 ft

D. Obtain the unsubmerged portion of the pipe by subtracting the submerged portion of the pipe circumference (C) from the total circumference of the pipe (A).

6.28 ft - 2.32 ft = 3.96 ft

E. Determine the unsubmerged area of the pipe that can receive the foam by multiplying the dry portion of the pipe (D) by the given length of the pipe (300 ft.).

3.96 ft x 300 ft = 1,188 sq ft can be foamed

F. Calculate the total cubic feet of foam required by multiplying the unsubmerged area that can receive the foam (E) by the desired thickness of the foam (in feet). We know that 3" of foam will be applied to the surface. Converting inches into feet, the desired thickness of the foam in feet is 0.25 ft.

1,188 sq ft x 0.25 ft = 297 cu ft

G. Convert cubic feet of foam into gallons of foam. A cubic foot is equal to 7.48 gallons.

297 cu ft x $\frac{7.48 \text{ gallons}}{1 \text{ cu ft}}$ = 2,221 gallons of foam

Calibrating Hose Retrieval Rate

To determine the correct hose retrieval rate, the operator must know: 1) how much foam is required per foot of sewer pipe, and 2) how much foam the application equipment is producing (in gallons per minute). The hose retrieval rate is determined by dividing the amount of foam produced per minute by the amount of foam required per foot. Refer to the label of the product being used for more specific information on hose retrieval rates.

This same procedure can be used when using surface coating techniques.

In the previous example, 2,221 gallons of foam is required to surface coat 300 feet of pipe with 3" of foam.

Since you know that 300 feet of pipe will be treated with foam, 7.4 gallons of foam is applied per foot (2,221 gallons of foam ÷ 300 feet).

Also, since you know that the equipment is generating 90 gallons of foam per minute, you can calculate the proper hose retrieval rate by dividing the amount of foam generated per minute by the gallons of foam applied per foot.

$$\frac{90 \text{ gallons of foam}}{\text{minute}} / \frac{7.4 \text{ gallons of foam}}{\text{foot}} = \frac{90}{7.4} = 12.16 \text{ feet/minute}$$

Determining the Effectiveness of Treatment

As with any pesticide treatment, always evaluate and determine the effectiveness of your treatment. Tracking the results of your treatments can point out deficiencies in your application methods and enable you to make better and more effective treatments in the future.

Conditions that can influence the effectiveness of a sewer root control treatment include:
- Improper application techniques, in particular, poor contact and exposure time,
- High sewer flows or surcharging conditions soon after application,
- Severe hydraulic sewer cleaning before or after treatment,
- Heavy grease deposits that interfere with pesticide contact, and
- Expired, ineffective, or improperly mixed pesticide.

Remember that treating roots with pesticides kills the roots but does not make them disappear. Several different decay organisms are constantly present in the sewer feeding on the dead roots. In addition, the buildup of solids and ever constant pressure caused by wastewater flows break the dead roots off, sending them to the treatment plant. This process may take weeks or even years. A few things to also remember include:
- If a sewer line is experiencing frequent blockage problems, treating roots with metam sodium will not immediately eliminate these blockages. You will first need to remove the blockages with a good cleaning (i.e., a mechanical method) before using pesticides. It is also important to remember that roots killed by metam sodium are not gone forever. The roots may regrow, meaning retreatments at 3 to 5 year intervals are often necessary.
- Live and dead root masses tend to look very similar during a video inspection. In time, if roots are treated effectively, they will decrease in size and break easily as an inspection camera passes.
- Another method of determining if treatment has been effective is by retrieving a root sample. When the bark layer of a sampled root is stripped and the root underneath is brown, that is an indication that the root has been killed. However, be aware that due to varying conditions within the sewer, the state of one root mass may not be representative of the root masses in the rest of the system.
- The most dependable method of determining if root control treatment has been successful is the decline of root related sewer problems. If there is a considerable decrease of problems a year after applying root control methods, then the root control treatment was successful.

Chapter 7.7 Review Questions

Correct answers are given on page 175.

1. Foam is used to deliver root control pesticides because it _____.
    a. breaks down rapidly after application
    b. effectively fills pipe voids
    c. can easily travel through pipes to manholes

2. Which type of glove is required if direct contact with metam sodium is likely?
    a. leather
    b. light-weight cotton
    c. chemical-resistant gauntlet type

3. Metam sodium is a _____.
    a. systemic herbicide
    b. non-selective contact herbicide
    c. pre-emergent herbicide

4. The desired foam consistency of metam sodium for sewer line root control use is that of _____.
    a. aerosol shaving cream bubbles
    b. foaming dish soap bubbles
    c. lathered shampoo bubbles

CHAPTER 7.8

Sanitizing Wooden Wine Barrels and Wine Bottle Corks

LEARNING OBJECTIVES

- ☑ Describe the purpose of sanitizing wine barrels and corks.
- ☑ Explain fumigation procedures to sanitize wine barrels and corks.
- ☑ Describe the formulation commonly used for sanitization of wine barrels and corks.
- ☑ Describe the PPE and air monitoring requirements related to fumigating wooden wine barrels and wine bottle corks.

Terms to Know
The following are important terms to know from this chapter. They are explained and *italicized* in the text and defined in the glossary at the end of this manual.

Antimicrobial   Eduction Tube

Sanitizing Wine Barrels and Corks

Purpose, Pests Controlled, and Fumigants Used

The fermentation of grape juice to produce wine requires the presence of naturally occurring yeasts and/or the addition of other specialized yeasts to the batch by the wine maker. Obtaining particular desirable qualities and flavor of the wine requires maintaining a delicate balance of these microorganisms. Certain other naturally occurring microorganisms can disrupt this balance during the fermentation process or after bottling the finished product. Winery personnel sanitize wooden wine barrels and wine bottle corks to minimize the effects of microorganisms that can affect the quality of wine but do not have a public health significance. Unless properly controlled, some of these organisms impart unpleasant flavors to wine, impact the aromatic qualities of the wine, or affect the wine's intended color. Others cause spoilage or interfere with the fermentation process.

Brettanomyces is a common yeast fungus that occurs on the skins of fruits in nature. High levels of this organism in grape juice used for the production of wine interfere with the flavor and taste of the wine batch. *Brettanomyces* or "Brett" can contaminate oak barrels used for fermentation and aging, requiring wine makers to sanitize the barrels before introducing new batches of grape juice. This organism easily spreads by using unsanitized equipment in wineries. Once the organism is present, it is difficult to eradicate. However, *Brettanomyces* is highly sensitive to sulfur dioxide (SO_2), an effective *antimicrobial* pesticide for suppressing the organism.

Several bacterial species, including *Lactobacillus* and *Pediococcus*, are contaminants that interfere with the wine making process. These organisms contribute to the production of excessive acetic acid, wine spoilage, and slowing the fermentation process (known as "stuck fermentation").

Wine barrels represent a sizeable investment for wine producers, making it crucial to reuse these barrels and extend their usefulness through proper sanitization methods. Although using SO_2 gas to fumigate barrels is the accepted standard in the industry, winemakers and researchers have experimented with various other methods, including:
- Steam cleaning the inner surfaces,
- Cleaning with ozone saturated water,
- Saturating the interior of the barrel with ozone gas,

- Using dry ice particles to blast the inner surfaces,
- Filling barrels with hot water,
- Burning sulfur wicks inside barrels.

Fumigation Practices

The practices involved in preparing for a wine barrel or wine cork fumigation with SO_2 gas have similarities to those already discussed. Refer to Table 4-1: Fumigant Practices in Non-Soil Fumigation (Chapter 4, page 57) for practices applicable to sanitizing wooden wine barrels and wine bottle corks.

Fumigating Wine Barrels with Sulfur Dioxide Gas

The gas cylinder containing the compressed SO_2 gas used for sanitizing wooden wine barrels must have an attached registered pesticide label. The product label must list sanitizing wine barrels as a registered use. Never use a compressed SO_2 product that does not indicate the use of the fumigant on wine barrels on the label.

Prewash Barrels. Wineries use various methods for washing out barrels that previously contained wine. These methods include using a high pressure water wash, soap, solvents, or other methods preferred by winery management. Once the barrels have been washed, they should be thoroughly drained and air dried according to the winery's procedures.

Read the Pesticide Label. Before fumigating wine barrels, read the product label, including the application information on it, and the company's operating procedures pertaining to this process. Understand all the hazards associated with using SO_2 gas and know how to avoid these hazards. Be sure you have the proper PPE required by the label.

Check to make sure that the cylinder you will use to fumigate wine barrels or corks contains SO_2 in the gas, not the liquid, form. Cylinders containing liquid SO_2 are under much higher pressure than those containing the gas phase. Using liquid SO_2 under high pressure will overdose the barrel and release too much SO_2, creating a serious respiratory hazard. Liquid containing cylinders employ an *eduction tube* to draw the liquid from the bottom of the cylinder, while the gas containing cylinders must not have this device. Some SO_2 labels caution against the use of a liquid SO_2 cylinder with an eduction tube as the high pressure in these tubes can damage barrels.

SO_2 Monitoring and Respiratory Protection

The general requirements and methods for fumigant safety monitoring were discussed in Chapters 3 and 4.

For SO_2, the work area must be monitored to know if and when a respirator is required. Various manufacturers produce air samplers that use detector tubes for SO_2 gas. These inexpensive samplers have a small hand pump or syringe to pull air from the atmosphere into the detector tube. You must wear the proper respiratory protection when taking air samples with this monitoring device. Be certain that the monitoring device is calibrated and can provide accurate readings at least in the range of 0.5 to 4.0 ppm SO_2. Know the limitations of these devices as other contaminants in the atmosphere can produce erroneous readings. You need to be properly trained on the use of these monitoring devices to reliably detect SO_2 level exceeding 2.0 ppm. If the concentration of SO_2 in the work area exceeds 2.0 ppm at any time, anyone working in or entering the fumigation area must wear the respiratory protection equipment specified on the label.

We discussed the PPE requirements for SO_2 in Chapter 5. If the concentration of SO_2 in the work area does not exceed the amount specified on the pesticide label, respiratory protection is not required. However, always have the proper respiratory equipment readily available and accessible in case of an emergency.

Fumigation Process

Before beginning the barrel fumigation process, you must post warning signs at all entrances to the area. The fumigant label will list the required information that must be included on the signs. Keep warning signs posted until after the application is complete and air monitoring verifies the ambient air concentration is below the safe and legal limit of 2.0 ppm SO_2. Do not allow anyone into the area without the appropriate respiratory protection until air monitoring confirms it is safe to enter.

Attach a barrel gassing unit with a pressure gauge to the SO_2 cylinder (obtain this equipment from the sulfur dioxide cylinder supplier) and open the cylinder valve. Be sure that the pressure gauge indicates that the cylinder pressure is within the range of 18 to 45 psi throughout the fumigation process. Most applicators prefer 20 psi. When the pressure drops below this level, connect a new cylinder to the gassing unit.

Begin treatment by removing the bung (stopper) from the cleaned barrel. Insert the SO_2 wand (gassing unit) into the barrel, then open the wand valve for 2 to 3 seconds. Immediately close the wand valve, remove the SO_2 probe, and seal the barrel opening with the bung. Repeat this process for each barrel you are fumigating.

During the fumigation process, regularly monitor the air in the work area to assure that the SO_2 concentration does not exceed 2.0 ppm. If readings are above 2.0 ppm, you and all other workers must leave the area immediately or put on the proper respiratory protection until the SO_2 level in the atmosphere of the work area drops to or below 2.0 ppm. The problem causing the unsafe concentration must be identified and corrected before resuming the fumigation.

Store the sealed treated barrels for at least one and no more than 30 days before filling them with grape juice or wine. Any barrels stored for more than 30 days must be retreated before use.

Fumigating Wine Bottle Corks

Fumigating corks prior to bottling and corking the bottles will reduce the effects of fungi or bacteria that may be present in/on the cork and that may contaminate the wine in the bottles. The process for fumigating corks is similar to the one for wine barrel fumigation. Use an SO_2 cylinder that contains gas, not liquid, SO_2. Be sure the cylinder provided by the supplier has a pressure gauge. Similar to the fumigation of wine barrels, the preferred pressure in the cylinder is about 20 psi. When the pressure drops below this level, use a fresh cylinder.

Follow the same procedures for posting the area where cork fumigation takes place as for barrel fumigation. Monitor the air in the fumigation area to assure that the SO_2 level stays below 2.0 ppm.

Place a plastic bag into the cork loading equipment and fill the bag with corks. Begin the plastic bag heat sealing process. Before completing the heat sealing, insert the SO_2 gas tube into the partially sealed bag and open the cylinder hose valve for 2 to 3 seconds. Remove the gas tube and complete the heat sealing. Place a "SO_2 Fumigated" label on the bag. Allow at least 24 hours to pass before using treated corks.

Other Fumigation Considerations

Review the following chapters for key concepts applicable to the fumigation of wine barrels and corks.
- Choosing the proper fumigant: Chapter 1, "Fumigant Basics."
- Site characteristics that influence a fumigation, air monitoring equipment: Chapter 3, "Planning for a Fumigation."
- Fumigant Management Plans (if required): Chapter 4, "Fumigation Safety."
- Securing the site, warning signs, protecting applicators, and the public: Chapter 4, "Fumigation Safety," and Chapter 5, "Personal Protective Equipment."

Chapter 7.8 Review Questions

Correct answers are given on page 175.

1. An effective antimicrobial fumigant for suppressing the fungus *Brettanomyces* that can contaminate wine oak barrels used for fermentation and aging is _____.
    a. hydrogen sulfide (H_2S)
    b. sulfur dioxide (SO_2)
    c. carbon monoxide (CO)

2. Wine corks that have been fumigated may be used after how many hours?
    a. 24 hours
    b. 56 hours
    c. 72 hours

CHAPTER 7.9

Remedial Wood Protection

LEARNING OBJECTIVES

- ☑ Differentiate between a) Sapwood and Heartwood and b) Softwood and Hardwood.
- ☑ List wood-destroying pests.
- ☑ Explain the purpose of remedial wood fumigations.
- ☑ Identify the fumigants used for remedial wood protection/preservation.
- ☑ Explain the application techniques and methods for remedial wood protection/preservation.
- ☑ Explain the environmental and public health concerns with remedial wood applications.
- ☑ Describe the factors that may influence the effectiveness of remedial wood treatment.

Terms to Know

The following are important terms to know from this chapter. They are explained and *italicized* in the text and defined in the glossary at the end of this manual.

Decay	Seasoning
Sapwood	Preservative
Heartwood	Check
Cellulose	

Remedial Protection of Wood with Fumigants

Pests Controlled
Wood protection fumigations control wood-destroying pests, including internal *decay* fungi and insects.

Purpose of Remedial Wood Fumigations
Wood decay is a major contributor to early failures of wood and wood-based materials in service. Remedial fumigation treatments are used to control wood-destroying pests to help protect in-service utility poles, pilings, fence posts, and other large timber members where other pest control methods are not as effective. After application, the fumigant, under vapor pressure, volatilizes and moves several feet from the point of application. Fumigants control existing decay fungi and sterilize the wood, including difficult to treat heartwood, to prevent the recolonization by decay fungi.

Before using fumigants for remedial wood protection, it is important to be properly trained and have an understanding of fumigation for remedial wood protection. An advantage of using fumigants is that they act quickly to destroy organisms. Fumigants are ideal for protecting heartwood and the inner regions of poles. Fumigants can also be effective in areas where other pest control methods have proven to be inadequate.

Fumigants Used
The primary fumigants used for remedial wood protection are MITC and the MITC-generators metam sodium and dazomet. These fumigants are applied at or near the ground line area in solid melt (MITC), liquid (metam sodium) or powder/granular (dazomet) forms. At the time of the manual's writing, some of these fumigants are labeled as Restricted Use Pesticides (RUPs), while others are not.

Fumigation Practices
The practices involved in preparing for a remedial wood protection fumigation treatment have similarities to those already discussed throughout this study manual. Refer to Table 4-1: Fumigant Practices in Non-Soil Fumigation (Chapter 4, page 57) for practices applicable to remedial wood protection.

Identification and Properties of Wood
Wood from a tree usually contains two sections: a lighter colored wood (*sapwood*) surrounding a darker colored wood (*heartwood*). Sapwood is the living, outermost portion, while heartwood is the dead inner wood which often comprises most of a cross section (Figure 1). As the older sapwood cells age and die, they cease to transport water or store energy reserves and become heartwood. A tree that has more sapwood than heartwood is indicative of a fast growth. Ninety percent (90%) of wood is composed of hollow fibers that carry nutrients and water, and the remaining 10% wood is made up of cells that transport food to the growing tissues between the bark and the wood. Under most conditions, sapwood is more susceptible to decay than heartwood.

Wood is also categorized as hardwood and softwood although these terms relate to their source and not to the actual hardness of the wood. Hardwood comes from deciduous trees such as oak, walnut, maple, birch, and mahogany. Hardwood trees do not have cones. On the other hand, softwood trees come from coniferous, or cone bearing, trees. Examples of coniferous trees are pine, cedar, fir, hemlock, redwood, spruce, and cypress. Most commercial wood comes from softwood trees.

Wood-Destroying Pests
Because wood is composed of *cellulose*, which is a sugar, it attracts many types of pests including fungi and insects. In addition, wood can become more vulnerable to pests when exposed to unfavorable conditions such as increased moisture or fluctuating temperatures. Fungi and insects are common pests that can be managed and controlled through remedial wood protection fumigations. Two invertebrate pests, carpenter ants and marine borers, are discussed in this chapter. Termites, beetles and fungi are discussed in Chapters 8.1, 8.2 and 8.4, respectively.

Carpenter Ants
While termites are the most damaging insects of wood structures, carpenter ants are the most destructive insects of wood products. They prefer to live in logs, trees, and stumps, but will damage any type of wood. Carpenter ants are so destructive that they can damage foam insulation to reach wood sources. These pests do not consume wood but use wood for housing and egg laying.

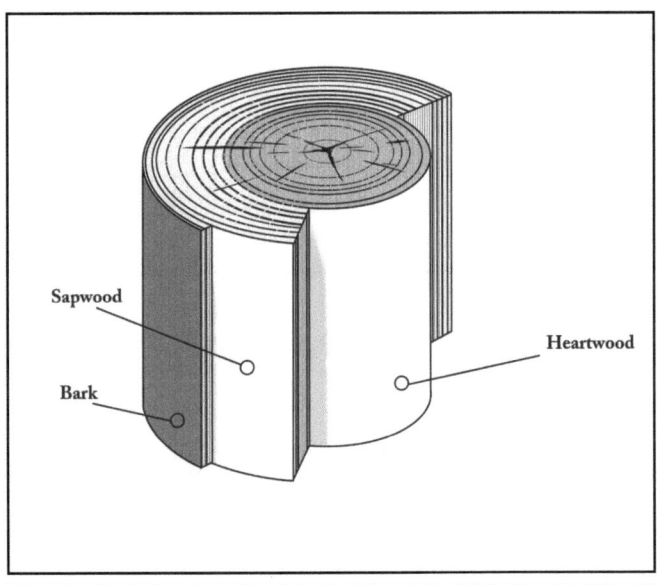

Figure 1: Transect of a piece of tree.

Marine Borers
Marine borers that damage wood include shipworms and crustaceans, such as gribbles. Shipworms enter and grow inside wood structures that are submerged in water. Because they enter wood products when they are small, it is difficult to notice there is a pest problem until significant damage has already occurred and the integrity of the wood structure is threatened. Gribbles, on the other hand, burrow just below the surface of wood. As a result, the wood becomes exposed to moisture that further deteriorates the wood. As the wood becomes more exposed, the pest burrows deeper into the next layer of wood.

Utility Pole Treatments
Cause of Utility Pole Deterioration
Utility poles are expensive to install and maintain. The investment in each new wood utility pole can vary depending on many factors, including the type and size of the pole and its location. Most utility companies maintain fairly precise records for each of their poles, both from an economic and a reliability standpoint, as well as from a safety standpoint, as linemen might be required to climb the poles at any time.

Utility poles deteriorate in place. The three major causes of pole deterioration are:

1. The *preservative* in the poles. The preservative may leach out, breakdown, or volatilize over time. As the preservative levels continue to drop, the active ingredient is reduced to a level where pests can reinvade the wood.

2. Persistence of wood damaging pests. Pests are always searching for new food sources. As soon as wood conditions are favorable, insect and fungal pests will invade the wood. Exposure to wind, sun, and rain may also play a role in making wood conditions favorable for pests.

3. Changes in moisture content of the wood. Changes in moisture can create favorable conditions for fungi and insects.

Utility poles are most prone to decay in the ground line zone (from 6" above grade to 18" below grade) and the pole top. The ground line zone is more vulnerable to wood-destroying pests because of the high moisture.

The decay hazard zone map (Figure 2) shows the severity of wood decay by geographic zones. There are five zones that range from one to five, with Zone five having the most severe decay. Parts of California are included in Zones 1, 3, and 4. However, microclimates can exist even within these zones. Care should be taken when inspecting utility poles in these microclimates. Carefully inspect utility poles in these microclimates as microclimates can lead to increases in fungal activity and pole deterioration. If decay not due to pests is already present (i.e., mechanical damage during storage), microclimates can lead to further decay that result in cracks or splits in the pole that serve as an entrance for pests.

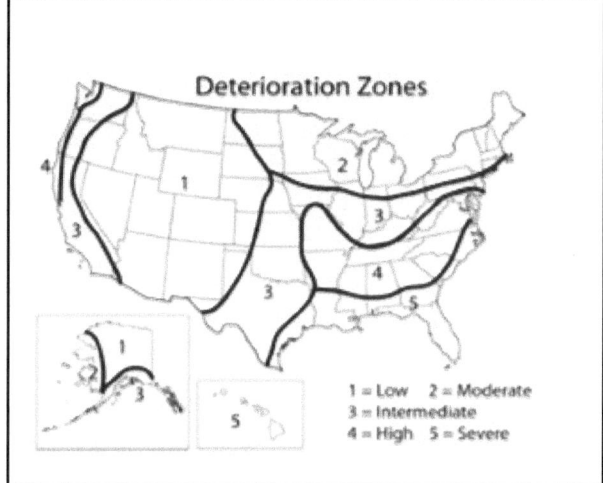

Figure 2: Decay Hazard Zones Map.

Reasons for Remedial Treatments

Pressure treated wood poles are not originally treated internally and the core may decay prematurely due to external mechanical damage, heavy splitting or early decay present before the pole was treated. With remedial treatments, the wood acquires a renewed level of protection.

Remedial treatment of utility poles, which is commonly practiced, is more cost effective than replacing the entire pole. On average, poles that are originally pressure treated but not retreated during their service life last for 25 to 40 years. Most companies will inspect in-service utility poles every 8 to 15 years for signs of deterioration. The life expectancy of these poles may be doubled if they are routinely inspected, and remedial treatments applied, when appropriate.

Remedial treatment reduces replacement costs by increasing the service life of the pole and reduces labor and equipment costs by decreasing the frequency of pole replacement. It is believed that by increasing the service life of utility poles by 50% or more, the demand for new poles and the stress on forest resources can be reduced by 33%. Finally, well maintained and sound poles can resist breakage caused by storms or changes to wire loads, making the poles safer, and resulting in fewer service interruptions.

Inspections

Utility companies will usually inspect in-service utility poles every 8 to 15 years. During inspection, a company determines the soundness of a pole and whether it can remain as is until the next inspection, or whether it should be replaced. Some companies use a method of retreatment in conjunction with inspection. After inspecting a pole both externally and internally, the inspection team methodically acts on the most beneficial option to the utility company. The options available to an inspection team are:

1. Emergency replacement - The pole is seriously weakened and in need of emergency replacement, or the pole is projected to be unsound through the next inspection period.

2. Remove decay (if possible) and treat with an appropriate pesticide for external decay, or fumigant for internal decay - Decay is present but the pole is projected to be sound until the next inspection.

3. No action - No decay is found. The pole is projected to be sound through the next inspection period.

This system gives utility companies a sound maintenance program that maximizes safety and reliability, as well as results in substantial long term financial savings. Figure 3 shows inspection and retreatment of utility poles.

Inspection and re-treatment

Photo 1. Soil around ground line is excavated to a depth of 18 inches and all loose dirt is removed from the pole with a wire brush.

Photo 2. An external preservative is applied to the ground line area of the pole with a long-handled brush. The preservative is liberally spread into the checks and cracks to a thickness of 1/16 inch

Photo 3. A wrapping or bandage is place over the treated area and stapled to the pole. The bandage usually consists of a plastic film and a backing of stiffening paper. The film acts as a moisture barrier between the soil and treated pole. The excavated area is backfilled and a date tag is affixed to the pole

Photo 4. Should a fumigant need to be applied, a hole is drilled, the fumigant is injected and the hole is plugged.

Figure 3: Inspection and retreatment of utility poles.

Application Treatments and Methods for Wood Preservation
Non-Fumigant Treatments

Wood preservation can include chemical and non-chemical controls. Some non-chemical control methods include limiting moisture content and the use of naturally pest resistant wood. The moisture content of wood plays a major role in its susceptibility to pests, including fungi. Methods that reduce or control moisture help preserve wood products. Drying wood can reduce moisture content. Utilizing wood that has natural defenses to pests can also be beneficial. Woods such as cedar and redwood are very resistant to pests. In contrast to limiting moisture content, timber stored for prolonged periods of time can be submerged in, or continuously sprayed with, water. This method can help reduce fungal growth by decreasing the oxygen required for fungal metabolism.

Sapwood of most commercial lumber tree species accepts preservatives much better than heartwood, and softwood species generally can be treated more uniformly than hardwood species. Preservative treatment by pressure is usually required for wood with a higher risk of attack by fungi, insects, or marine borers.

The three exposure categories for preservatives are:

1. Ground contact (high decay hazard that needs a heavy-duty preservative),

2. Above ground contact (low decay hazard that does not usually require pressure treatment), and

3. Marine exposure (high decay hazard that needs a heavy-duty preservative or possibly dual treatment).

Pesticides used for wood-destroying pests include fumigants as well as various wood preservatives (i.e., waterborne and oil-borne protectants). Water-borne protectants include: chromated copper arsenate type C (CCA-C), copper azole type B (CA-B), copper azole type C (CA-C), alkaline copper quaternary (ACQ), and disodium octaborate tetrahydrate (DOT), as well as applications of other fungicides (i.e., propiconazole and tebuconazole) or insecticides (i.e., permethrin, imidacloprid, or chlorothalonil). Oil-borne protectants include copper naphthenate (CuN) and active ingredients in the Isothiazolinoe class. Depending on the product, these pesticides may be applied by various methods including pressure treatment, brushing, spraying, pouring, dipping, cold soaking, thermal processes, double diffusion, and others. In some cases, the soil around a utility pole is removed before the preservative, like borate or copper, is applied and the site wrapped with a preservative pad or bandage. This study manual will focus on the fumigants used for remedial wood treatment.

Fumigant Treatments

Fumigants are very effective in protecting the integrity of wood and preventing further damage by wood-destroying pests. This is because fumigants can more easily reach pests compared to other methods that fail to adequately reach the pest, and result in the ineffective control or eradication of the pest. Moreover, visual observation may not indicate the severity of a pest problem and pests can go undetected. A fumigant will provide a more adequate and efficient treatment since it can reach pests that are not easily visible. Lumber, utility poles, and other wooden material can be treated with fumigants during a commodity fumigation (Chapter 7.2). This chapter focuses on remedial fumigation of in-service utility poles, pilings, and similar large timber members. The fumigants used for remedial wood protection are MITC and the MITC-generators dazomet and metam sodium.

Logs, timber, and wooden materials known to have internal decay should be scheduled for treatment before the extent of decay has reached a point that requires replacement. For remedial treatment of wooden poles/timbers:
- Do not treat structures/beams indoors.
- Do not drill an application hole through *seasoning* checks to apply product. If the hole intersects a *check*, plug the hole and drill another. If more than two treatment holes intersect an internal void or rot pocket, redrill the holes farther up the pole into relatively solid wood.

Remember to always read the product label and follow correct application procedures. Generally, applicators first drill treatment holes into the wood at a downward 45 degree angle (or greater) to a depth of between 14 to 16 inches (check the label for specific requirements). For wood in ground contact, the first hole should start at or slightly below ground line. Generally, holes in the treatment zone should be drilled in a spiral pattern with the next hole generally rotating 90 degrees from the previous one with a 6 to 12 inch vertical space between the holes. The applicator, wearing proper PPE, then places the fumigant (MITC, metam sodium, or dazomet) inside the drilled hole. The application rate is based on pole circumference or a certain amount of fumigant per hole. The treatment holes should not be overfilled to allow room for a plug. Finally, the hole is immediately plugged with a cork or similar device after the application.

With dazomet, an accelerant of a 1% solution of copper naphthenate in mineral spirits may be added to treatment holes after application of the product to speed up the breakdown of the active ingredient and release of the active fumigant inside the wood product. Keep the product away from the accelerant until it is added to the treatment holes and immediately plug the hole.

Application Concerns

The fumigants used as remedial wood protectants to control fungi and insects that destroy wood have to be inherently toxic to be effective. There are environmental and human health concerns associated with the use of these fumigants. While these fumigants are important in protecting wood products, they are not selective and can harm non-target organisms. The fumigant product label may indicate that the fumigant is toxic to one or more organisms such as mammals, birds, fish, aquatic invertebrates, oysters, shrimp, or terrestrial or aquatic plants. Always follow the guidelines for the use, storage, disposal, and spill cleanup of these products to mitigate harm to non-target organisms or the public. Environmental contamination from these pesticide products can occur from runoff into storm drains, lakes, rivers, streams, and in soil environments and can harm fish and other wildlife. Improper disposal, spills, and leaching of wood preservatives can also affect groundwater sources. Always follow the product label on the proper disposal of the product to reduce environmental contamination.

Human health effects from exposure to these pesticides can either be acute (short-term) or chronic (long-term) and are determined by exposure time and route of exposure. Make sure applicators are properly trained in the use of fumigant products for remedial wood protection and that they wear the required PPE.

Effectiveness of Treatment

The effectiveness of a fumigant treatment depends on the proper use of product, including the correct penetration depth of the preservative. Even if a fumigant can easily reach a pest, effective control is not achieved unless the fumigant was correctly applied. Proper application promotes longer lasting control.

Regularly monitor treated wood for signs of damage from pests as retreatment may be necessary over time.

Other Fumigation Considerations

Review the following chapters for key concepts related to remedial wood fumigation.
- Choosing the proper fumigant: Chapter 1, "Fumigant Basics."
- Site characteristics that influence a fumigation, air monitoring equipment: Chapter 3, "Planning for a Fumigation."
- Fumigant Management Plans: Chapter 4, "Fumigation Safety."
- Securing the site, warning signs, protecting applicators, and the public: Chapter 4, "Fumigation Safety," and Chapter 5, "Personal Protective Equipment."
- Application equipment and aerating the site: Chapter 6, "Fumigation Methods and Application Equipment."

Chapter 7.9 Review Questions

Correct answers are given on page 175.

1. When conducting remedial wood protection, it is important to know that under most conditions, heartwood is _____.
    a. less susceptible to decay than sapwood
    b. as susceptible to decay as softwood
    c. more susceptible decay than redwood

2. Which of the following is a desired outcome of remedial wood treatment of utility poles?
    a. increased serviceable life of the pole
    b. increased frequency of pole replacement
    c. increased presence of beneficial pests

3. The probable target pest in remedial wood protection is _____.
    a. bacteria
    b. virus
    c. fungi

CHAPTER 8.1

Termites

> **LEARNING OBJECTIVES**
>
> ☑ Discuss the importance of distinguishing between termites and ants.
> ☑ Describe the three types of termites and the termite life cycle.
> ☑ List the fumigants used to control termites.
> ☑ Discuss pest factors that influence the suitability of fumigation as a control method for termites.

> **Terms to Know**
> The following are important terms to know from this chapter. They are explained and *italicized* in the text and defined in the glossary at the end of this manual.
>
> Frass

Termites

Termites are among the most successful groups of insects on Earth. They are sometimes called "white ants," although they are not closely related to ants (they are more closely related to cockroaches). Termites, like ants, are social insects. This means they live in a large group or colony where different members (e.g., workers, soldiers, and reproductives) perform different tasks. There are over 3,000 named termite species around the world, with about 50 species present in the United States. Depending on species, termites might have small colonies of only a few thousand individuals (e.g., drywood termites) while others can have colonies reaching up to a million individuals (e.g., subterranean termites). Not all termite species are destructive to structures, but the ones that are destructive can cause extensive damage.

The destruction termites cause is related to their ability to digest wood and turn it into food. They are able to do this because of the special protozoans and bacteria that live in their digestive tract that enable them to digest cellulose and other products within wood. In nature, this ability plays a crucial role in converting dead wood into useful organic matter that other plants and animals can use. It also keeps the landscape from being littered with fallen logs and wood.

In California, the Structural Pest Control Board issues the license for the use of fumigants, such as sulfuryl fluoride, to control structural pests (like termites) in structures. This study guide does not cover these uses. Applicators fumigating commodities, like timber, for termites must hold the DPR Non-Soil Fumigation category. There are also some items, like lumber or furniture, which can be fumigated for termites under either license.

Termite Biology

Knowing how to differentiate termites from ants and having knowledge of termite life cycle, biology, and behavior are essential in termite control. This information serves as the basis for determining the best control methods to use. For some species of termites, and in specific infestations, fumigation may not be the best control option.

Termite Social Structure

Termites are social insects that live in colonies. Some termite species live in the ground, others live inside wood. Each colony has a highly developed and complex social organization called a "caste" system where immature forms of termites can develop into adult forms that perform different functions within the colony (Figure 1).

Workers

"Worker" termites make up most of a colony. They are white or cream-colored, soft-bodied and about 1/4 of an inch long (Figure 2 (B)). They have no eyes or wings, and species differ in their sensitivity to light and dryness.

Figure 1: Termite life cycle. Termites undergo simple metamorphosis where immature stages are morphologically similar to adults. The life cycle of termites does not have a pupal stage. Termites have a caste system where an immature form develops into an adult form with a specific role in the colony (e.g., workers, soldiers). (Illustration © Jeanie Tomasko)

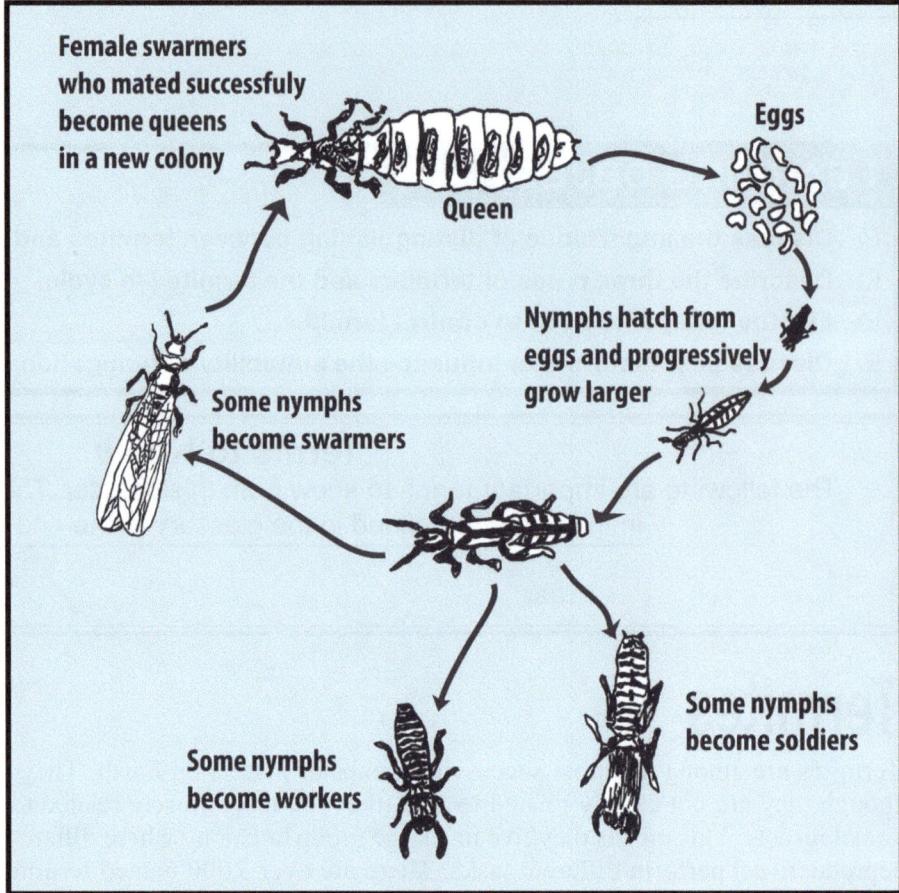

Workers are sterile and perform the upkeep and maintenance of the colony and are responsible for feeding the queen. Both nymphs and adult workers have thin, bead like antennae and differ only in size.

Soldiers

In some colonies, some nymphs develop into wingless, light colored termites with large brown heads with well developed dark brown jaws (Figure 2 (A)). Referred to "soldiers", these termites defend the colony from attack by ants or other termites. The sterile soldiers are far less numerous than the workers and are not seen unless the nest is disturbed.

Reproductives or Swarmers

As a colony matures, some nymphs develop wing pads and at the final molt turn dark and emerge as winged adults (Figure 3). These are the future "swarmers," also called primary "reproductives" or "alates," which is Latin for "having wings."

Reproductives or swarmers are weak flyers. They are dark brown or black, flattened, and about 3/8 of an inch long

Figure 2: Termite workers. "Workers" surrounding a "soldier" with a dark head at bottom (A). A single termite worker (B). A shelter tube (C). Some subterranean termite species are more dependent on moisture and will build shelter tubes to travel between the nest and their food source. Here, the tube has been broken open revealing the workers underneath. Drywood termites do not build these tubes. (Photo credits: (A and B) © Gary Alpert, Harvard University; (C) © Whitney Cranshaw, Colorado State University. All from Bugwood.org).

with big compound eyes. Only a small number of the alates survive to establish their own colonies. Most reproductives will fall prey to predators such as birds, toads, reptiles, and insects (primarily ants). Many others die from dehydration or injury. After a reproductive takes flight and finds a mate, it will lose its wings and will search for a place to start a nest, mate, and begin rearing the first group of workers. The mated female becomes the "queen" (Figure 4), and the male the queen mated with, the "king." The king remains with the queen since periodic mating is required for continued egg laying.

Secondary Reproductives

In very large colonies, nymphs may develop into secondary reproductives, sexually mature adults with reduced wings (Figure 5). These secondary reproductives can, if needed, take over the entire egg laying needs of the colony. Strong colonies can have multiple queens. Secondary reproductives also give the colony a chance to spread through the process of "budding." In budding, several workers and secondary reproductives cut off from the main colony and form a new, self sufficient colony.

Ants vs. Termites

Sometimes people mistakenly refer to termites as "white ants." While they do share some similarities, you should know how to differentiate between ants and termites because the methods used to manage and control each species will differ. Both ants and termites are social insects and live in colonies. Superficially, they may look similar. However, their life cycles are different. Termites undergo simple metamorphosis, whereas ants undergo complete metamorphosis.

Figure 6 illustrates how an examination of the antenna, waist, and wings of these two insects can help you easily differentiate them.

Pest Identification and This Study Manual

This study manual will cover the three common termite types. This study manual is not intended to be a definitive pest identification manual. Proper identification of the termite species is essential in pest management and will help you decide on the best control or management method (i.e., fumigation), the pesticide active ingredient(s) to use, the correct dosage, etc. If necessary, consult an expert (i.e., entomologist, biologist, extension worker) to help with pest identification. Always read and follow the pesticide label for use information.

Types of Termites

It is useful to group termites into three types based on the habitat they prefer:
1. Drywood,
2. Dampwood, and
3. Subterranean termites.

Figure 3: Reproductives or Swarmers. Large numbers of winged termites emerge on warm sunny days after a rain in late spring. Called "swarming," this collection of winged termites is often the homeowner's first indication of an infestation. (Photo © Susan Ellis, Bugwood.org).

Figure 4: Termite castes. The queen termite is much larger than the workers and the soldiers. Her distended abdomen is an egg laying machine, producing 5,000 to 10,000 eggs per year. (Photo © Lyle Buss, Entomology & Nematology Dept., University of Florida).

Figure 5: Secondary reproductives. Secondary reproductives have wing pads instead of fully developed wings. Unlike primary reproductives that fly away to mate and find a new colony, secondary reproductives mate and reproduce within the existing colony. (Photo © Phil Sloderbeck, Bugwood.org).

Drywood termites build their nests in sound, dry wood above ground and are typically found in hardwood forests. They sometimes infest lumber as well as dead limbs of trees, utility poles, posts, and other exposed wood. They also damage furniture, picture frames, and other wooden items in storage. They have a low moisture requirement and can tolerate dry conditions for prolonged periods. They remain entirely above ground and do not connect their nests to the soil. All the water they need to survive comes from the wood they eat, and the water produced within their cells. Drywood termites eat both the softer spring and denser summer growth rings of wood, often leaving the infested wood riddled throughout with galleries (tunnels that run with the grain of wood). Drywood termites can also cut across the grain in wood. Piles of their fecal pellets (*frass*) may be a sign of their presence. The termites often push frass out of the infested wood through small holes. The drywood termite pellets are cylindrical with six depressions on the sides. Due to this unique shape, the pellets can be used for identification.

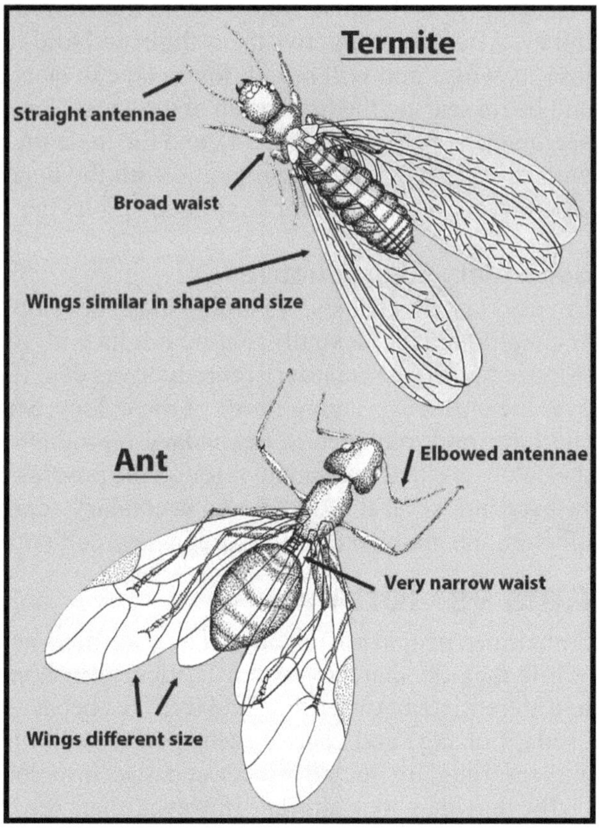

Dampwood termites build their nests in moist, decaying wood but can extend tunnels into drier parts of the wood. They are typically found in coniferous forests. Dampwood termites are larger than either the drywood or the subterranean termite. They typically infest posts, forest litter, dead trees, and wooden structural beams in soil or areas with sufficient access to water. Because of their need for water, a dampwood termite infestation is usually a sign of a structural fault such as a leaky roof or plumbing, condensation build up, or similar high humidity situations. In California, these types of repairs are generally performed under a Branch 3 license from the Structural Pest Control Board.

Subterranean termites, as their name implies, generally live underground, and, with rare exceptions, must maintain contact with the soil. They work tunnels through soil to reach wood above ground. They live in diverse environments, from deserts to savannahs to forests. In the U.S., subterranean termites are the most widespread and of the three termite pests and cause the most damage. Shelter tubes (or mud tubes) containing live workers are a hallmark sign of an active infestation. They are found in every state except Alaska. Besides eating fallen wood, they may also feed on wooden structures such as buildings, telephone poles, and fence posts.

Figure 6: Differentiating termites and ants. There are physical differences between the winged forms of termites and ants. In both insects, only the reproductive forms have wings, and the majority of the colony are without wings. However, the same physical differences illustrated above also apply to the non-winged forms (e.g., workers, soldiers) of both insects. (Illustration © USDA Forest Service, Bugwood.org).

Fumigation For Control of Termites

Use of fumigants, non-fumigant insecticides, or non-chemical means to control termites largely depends on the type of termite causing damage and what they are damaging. Prior to an application, ensure the control method (e.g., fumigant, non-fumigant) is appropriate for the pest you are trying to control.

Drywood termites: Fumigants will most likely successfully control drywood termites since they do not depend on access to a water source, and more typically nest in wood. Chamber, commodity, or spot fumigation can be performed for infested items such as logs, lumber, or furniture under a non-soil fumigation license from DPR. Fumigation of the whole structure to control termites is usually the only practical and effective method to treat areas of high infestation. In California, fumigations to control structural pests must be performed under a Branch 1 license from the Structural Pest Control Board.

Dampwood termites: It is seldom necessary to use pesticides to treat an infestation of dampwood termites. These insects depend on a water source. Their presence is usually indicative of a structural problem in the building that allows them access to a water source (e.g., a leaking pipe that dampens wood, a leaky roof). If dampwood termites are causing a structural problem, removing their source of the water through structural repairs (i.e., repairing a leaking roof or pipe) will also remove the moisture they need to survive. Structural repairs that eliminate the conditions that impact wood will make the structure an unfavorable habitat for the pest. In California, these types of repairs are generally performed under a Branch 3 license from the Structural Pest Control Board.

Subterranean termites: Often, fumigants are not effective against subterranean termites because they almost always nest underground. The most common methods to control subterranean termites that do have underground nests are to take measures that prevent them from establishing colonies, or use of liquid pesticide treatments and baiting systems. In California, these types of treatments are generally performed under a Branch 3 license from the Structural Pest Control Board.

Fumigants for Termite Control

Sulfuryl fluoride and methyl bromide are labeled for termites. The use of methyl bromide for termite control is limited to non-inhabited uses for quarantine fumigation of wood products. When fumigants are used, tarps are typically used to confine the fumigant and allow gas to penetrate thoroughly.

Chapter 8.1 Review Questions

Correct answers are given on page 175.

1. Which of the following best describes the life cycle of a termite?
    a. simple metamorphosis
    b. complex metamorphosis
    c. ametabolous

2. Which of the following fumigants is labeled for termites?
    a. carbon dioxide
    b. aluminum phosphide
    c. methyl bromide

3. Which of the following makes up most of a termite colony?
    a. workers
    b. swarmers
    c. secondary reproductives

CHAPTER 8.2

Wood-Boring Beetles

> **LEARNING OBJECTIVES**
> - ☑ Discuss the importance of properly identifying the species of wood-boring beetle prior to control.
> - ☑ List the three groups of wood-boring beetles and how to differentiate between the groups.
> - ☑ List and discuss the pest factors that influence the suitability of fumigation as a control method for wood-boring beetles.

Wood-Boring Beetles

There are many species of beetles that infest and damage wood. Some species of beetles feed only on dead or dying trees and cannot reinfest sound, seasoned lumber. Others destroy wooden objects by repeatedly attacking and breeding in the wood.

You will seldom see the larvae and adults of wood-infesting beetles. When wood damage is found, you must examine the wood itself and use emergence holes, the type of frass present, and the age and species of wood as clues to identify the insect responsible for the damage. Because control methods and potential wood damage differ amongst wood-damaging beetle species, accurate identification is essential to effective control. Even the most aggressive wood-destroying beetle works slowly so there is time to accurately identify the pest and consider the treatment alternatives.

In California, the Structural Pest Control Board issues the license for the use of fumigants, such as sulfuryl fluoride, to control structural pests (like powderpost beetles) in structures. This study manual does not cover these uses. Applicators fumigating commodities, like timber, for wood-boring beetles must hold the DPR Non-Soil Fumigation category. Note there are some commodities, like lumber, which can be fumigated for wood-boring beetles under either license.

Pest Identification and This Study Manual

This study manual will cover some of the common wood-boring beetle species you might encounter in your work. It is not an exhaustive discussion and is not meant to be a definitive pest identification manual. Accurate identification of wood-boring species is essential in pest management and will help you determine the best management or control method (i.e., fumigation), the pesticide active ingredient(s) to use, the correct dosage, etc. If necessary, consult an expert (i.e., entomologist, biologist, extension worker) to help with pest identification. Always read and follow the pesticide label for use information.

Powderpost Beetles

Powderpost beetles are second only to termites in their economic importance. Their name refers to the extremely fine, flour-like frass that the emerging adults push out of infested wood (Figure 1). Because the powder is loose in the feeding tunnels, it may continue to sift out long after the adults have emerged and even after the infestation has died out.

Figure 1: Powderpost beetle frass. Loose powder is often the first noticeable sign of a powderpost beetle problem. (Photo © Stan Lebow, USDA Forest Service).

"Powderpost beetles" is used to collectively refer to three groups of beetles: lyctids, anobiids, and bostrichids. All three groups can infest sound lumber and collectively, can attack both softwood and hardwood. Although many infestations are determined solely by the presence of emergence holes and are referred to as being from the "powderpost beetle complex," it is important to be able to identify and differentiate between these groups.

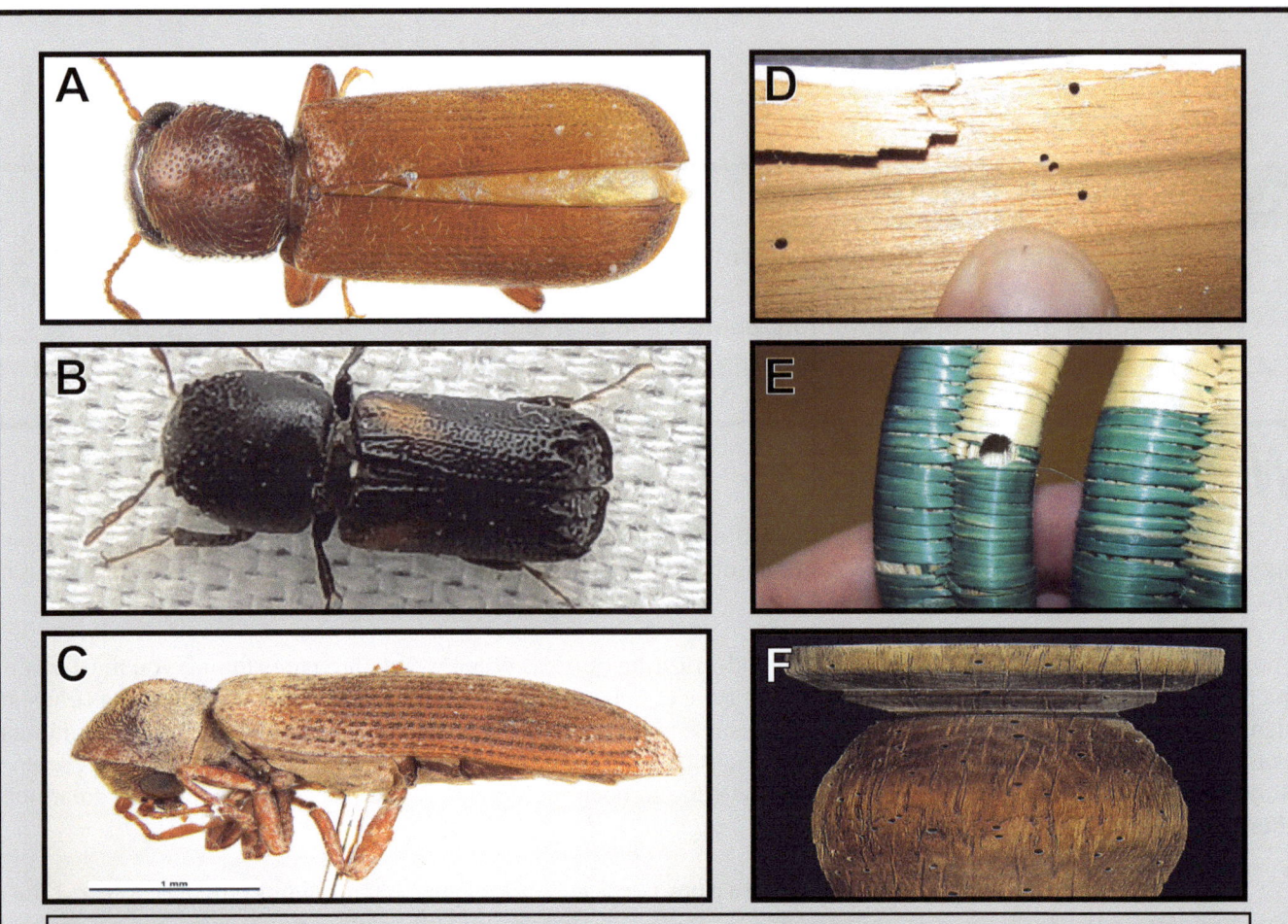

Figure 2: Powderpost beetles. Lyctids (A) are often called "true powderpost beetles" while Bostrichids (B) are sometimes called "false powderpost beetles." Anobiids (C) are another group of powderpost beetles. Although all three groups cause similar damage, they differ in appearance, size of exit holes they produce, frass characteristics, and life cycles. Pictures D to F illustrate damage caused by powderpost beetles. (Photo credits: (A and C) © Pest and Diseases Image Library, Bugwood.org; (B) © Anita Gould, iNaturalist.org; (D and E) © UW-Madison Dept. of Entomology; (F) © Gyorgy Csoka, Hungary Forest Research Institute, Bugwood.org).

"True" Powderpost Beetles (Lyctids)

Adult lyctids resemble the red flour beetle (Figure 2-A). They are small (1/12 to 1/5 of an inch long), elongated, flattened, reddish brown to black, and their head is visible when viewed from above. They have short, 11 segmented antennae with a 2 segmented terminal club. Adult females lay eggs in the surface pores of wood. Larvae hatch and tunnel through the sapwood, usually along the grain. When they complete their development, the adults emerge from the wood through small (1/32 to 1/16 inch) holes that are the telltale signs of powderpost beetle damage (Figure 2-D). Adults are attracted to light and often appear at windowsills in infested rooms.

True powderpost beetles attack only dry (8% to 20% moisture), seasoned hardwoods and prefer porous wood such as ash, oak, mahogany, hickory, maple, walnut, bamboo, and wicker. Most infestations occur in sapwood, and they use the wood as both food and shelter. The insects are generally brought into buildings in wood that contains their eggs or larvae. It will normally take years for serious structural damage to occur from lyctids.

"False" Powderpost Beetles (Bostrichids)

Bostrichids are most abundant in the tropics, and in the U.S. do not present as much of a pest problem as lyctids and anobiids. Adult false powderpost beetles are dark brown to black and have a cylindrical shaped and stout body (Figure 2-B). Adults range from 1/4 to 2 inches long.

Bostrichid frass contains small bits of wood, a few pellets, and fine powder. The frass feels gritty and tends to clump together. Damage from false powderpost beetles is typical of other wood-boring beetles. Adults make round exit holes as they emerge from wood. Unlike powderpost beetles and anobiids, female bostrichids bore into the wood to lay their eggs. False powderpost beetles attack both soft and hardwoods, but they prefer the latter.

They infest wood that has a moisture content of 6% to 10% and generally like newer versus aged wood. Unlike true powderpost beetles and anobiids, false powderpost beetles do not usually reinfest the same piece of wood after it has become dry and seasoned.

When bostrichids are found, it's usually because they came with hardwoods shipped from other countries. For example, the bamboo powderpost beetle is found in baskets, picture frames, furniture, and other imported bamboo material (Figure 2-E).

Anobiid Powderpost Beetles

Anobiid powderpost beetles (i.e., the furniture or deathwatch beetle) range from 1/25 to 1/3 of an inch long and are usually reddish brown, brown, or gray (Figure 2-C). They have a distinctive hood like, bell shaped thorax, which conceals their head. Unlike the true powderpost beetles, you cannot see the head of an anobiid from above. Anobiid powderpost beetles do not have a club on the end of their antenna, but the last 3 segments of their antennae are usually longer and broader than the other 8 segments. They resemble the cigarette and drugstore beetles that are stored commodity pests discussed in the next chapter.

The obvious sign of an infestation of anobiid powderpost beetles is the accumulation of powdery frass and tiny pellets underneath infested wood. Fresh powder is bright and light colored, similar to freshly sawed wood. Small (1/25 of an inch long) oval pellets fall from infested channels or old emergence holes. These pellets give the frass a gritty quality. Another sign of anobiid powderpost beetles is tightly packed frass in the tunnels. If frass is yellowed and partially caked on the surface where it lies, the infestation has died out. Emergence holes are round and from 1/16 to 1/8 of an inch in diameter.

Anobiids attack the sapwood of both hardwoods and softwoods and the infestation can extend into the heartwood. It normally takes 10 or more years for the beetles to increase in number before an infestation is noticeable. Adults prefer high moisture conditions which allows for shorter generation times. Wood with a moisture content of 13% to 30% is most susceptible to attack. Infestations often start in poorly heated or unventilated crawl spaces and spread to other parts of the house. Old unheated buildings, livestock housing, and homes with earthen basements or damp crawl spaces are prime targets for an anobiid infestation. Anobiids are much more likely than other species of wood-infesting beetles to infest old lumber and wood. Old barn board used as paneling is often a source of infestation.

Powderpost Beetle Management

The first task in managing a powderpost beetle problem is to determine if the infestation is still active. The presence of adult beetles, new emergence holes, and powder like frass are indicative of an active infestation. It is sometimes useful to mark existing emergence holes and check for the appearance of new holes after 3 to 6 months. Placing a dark colored cloth under infested areas and checking for piles of powder over a period of time will also help verify activity. If part of the wood can be sacrificed, you can remove small sections of the wood to look for adults and larvae.

Fumigation can be used in situations where liquid treatments might not be possible because infested items are stored on pallets. Fumigation is also useful for wooden containers (e.g., barrels) used for storing food where other treatments (such as borates) would not be appropriate.

Fumigation could be justified if an infestation is widespread. Fumigation may also be warranted if the infestation has spread into areas where access to, or removal of, wood is impractical. However, remember that while fumigation will help control an existing infestation, it will do nothing to protect the wood from future attack. Fumigation of infested furniture, antiques, and other manufactured articles can be done at a substantially lower cost by fumigating the items under tarps, in trailers, or in chambers.

Chapter 8.2 Review Questions

Correct answers are given on page 175.

1. Which of the following is true of fumigations to control wood-destroying beetles?
    a. the primary fumigant used to control wood-destroying beetles is sulfur dioxide
    b. fumigation will protect wood from future attack for up to 5 years
    c. fumigation will help control an existing infestation, but cannot protect wood from future attack

2. "Powderpost beetles" is used collectively to refer to which three groups of beetles?
    a. lyctids, anobiids, and bostrichids
    b. diptera, tabanidae, and muscidae
    c. aedes, culicidae, and anophelinae

3. Which of the following generally prefers to infest newer wood with a moisture content between 6 to 10%?
    a. true powderpost beetle
    b. false powderpost beetle
    c. anobiid powderpost beetle

4. The first step to managing a powderpost beetle infestation is to _____.
    a. determine if the infestation is still active
    b. apply a broad spectrum insecticide
    c. remove the beetle's food source

CHAPTER 8.3

Stored Commodity Pests

> **LEARNING OBJECTIVES**
> - ☑ Discuss the importance of accurately identifying stored commodity pests.
> - ☑ Identify important stored commodity pests.
> - ☑ Differentiate between primary and secondary feeders.
> - ☑ Discuss the economic significance of stored commodity pests.
> - ☑ List and discuss the pest factors that influence the suitability of fumigation as a control method for stored commodity pests.

Pests of Stored Products

A large number of beetles, several moths, psocids (known as "booklice"), and some mites commonly infest various food and non-food products. In general, these pests feed on a variety of plant material, including spices, flour, vegetable seeds, tea, dried flower arrangements, dried herbs and fruits, chocolate, pet food, grain, tobacco, some rodent baits, straw, bamboo, cigars, and nuts. Some pests also feed on materials of animal origin, such as dried fish and meats, milk powder, and non-food materials like hides, skins, wool, and other fibers. Under optimal conditions (temperature and humidity) these pests can develop and reproduce continually.

Primary and Secondary Feeders

Insect pests of stored agricultural products fall under two groups: primary and secondary feeders. You can find both primary and secondary feeders living in grain storage areas at the same time.

Primary Feeders

Primary feeders can destroy whole, sound grain. These insects feed on the grain, reducing its nutritional value and its ability to sprout. Damage caused by primary feeders makes grain more susceptible to secondary feeders (insects that feed on grain debris). Adult females of primary feeders deposit eggs on or in whole kernels. Larvae develop within the kernels. Some primary insect feeders include the:
- Angoumois grain moth,
- Rice, granary, and maize weevils, and
- Lesser grain borer.

Secondary Feeders

Secondary feeders feed only on damaged grains and seeds. The outer layer of the grain or seed must be damaged (i.e., cracked, holed, abraded, or broken) for the pest to gain access. This damage can occur during harvest or be caused by rapid drying, processing, or feeding by a primary feeder. Some common secondary feeders of raw agricultural products include the:
- Indian meal moth,
- Mediterranean flour moth,
- Tobacco moth,
- Drugstore and cigarette beetles,
- Confused and red flour beetles,
- Dermestid beetle,
- Sawtoothed and merchant grain beetles,
- Yellow mealworm beetle, and
- Grain mites.

Pest Identification and This Study Manual

This chapter will cover some of the common stored commodity pests you might encounter in your work. It is not an exhaustive list, nor is it meant to be a definitive pest identification manual. Accurate identification of stored commodity pest species is essential in pest management and will help you determine the best management or control method (i.e., fumigation), the pesticide active ingredient(s) to use, the correct dosage, etc. If necessary, consult an expert (i.e., entomologist, biologist, extension worker) to help with pest identification. Always read and follow the pesticide label for use information.

Moth Pests

Moths are second only to beetles in the amount of damage they cause to stored products. Moths go through complete metamorphosis. Only the larval stage causes damage.

General Moth Characteristics

Adult moths have 2 pairs of wings that fold over the body when the moth is resting. Like butterflies, some moths have color patterns on their wings while others have solid colored wings. While wing color and patterns can be helpful in identification, wing scales can rub off making identification difficult. The antennae of female moths are long and slender. The male's antennae are often long and feather like but this can vary between species. Adults live for a short time, breed, and do not feed on grains. Females die soon after they lay their eggs.

Moth larvae, called caterpillars, resemble small worms with legs. Some species have distinct color patterns that can help with identification. It is easy to confuse moth larvae with beetle larvae. An easy way to tell the difference is to look at the middle of their bodies. Moth larvae have 6 legs at the area just behind the head and often fleshy, leg like appendages called "prolegs" in the middle of the abdomen.

Moths can produce several generations per year, depending on the temperature and food source. In favorable conditions (warm temperatures and abundant food), they may complete their life cycle in one to three months.

Moth Damage

Moth larvae eat and contaminate grain making it unfit for human consumption. Some moths infest whole grains, while others infest milled or ground foods such as flour, cereals, and pet food. Damage usually occurs when these items are stored for an extended period. If control is inadequate, moths may follow a product throughout the manufacturing and distribution process. You can find moth pests in fields, storage bins, mills, delivery trucks, retail stores, and homes.

Indian Meal Moth (*Plodia interpunctella*)

Indian meal moths are one of the most common pests of grain storage and processing facilities as well as stored products in homes and warehouses (Figure 1-A). Larvae (Figure 1-B) often migrate away from the food source to spin a cocoon in bag seams, cracks and crevices, or other sheltered sites (Figure 1-C). Indian meal moths can produce 5 to 6 generations a year under ideal conditions (86°F and 70% relative humidity).

Indian meal moths can cause major infestations in food manufacturing plants or warehouses. There is a low tolerance for moths because of the potential for food product contamination. Pheromone traps can be used in warehouses to detect initial moth activity and for trapping to isolate and identify breeding sites.

Indian meal moth larvae feed a few inches from the surface of the grain mass and "web" the grain together. If populations are high, the surface will become crusted, which can protect the larvae from surface applied insecticides or fumigants. Remove the crust and damaged

Figure 1: Indian Meal Moth.
(A) Adult Indian meal moths are 1/3 of an inch long. Their outer wings are two-toned: a tan to gray wing base with a coppery reddish or bronze outer two thirds. (B) Larvae are cream colored, 1/2 inch long caterpillars. The caterpillars have chewing mouthparts and feed on a large variety of commodities including stored grain, nuts, rice, dog biscuits, dried fruit, bird seed, and much more. (C) The caterpillars can spin large amounts of silk webbing in and over food. (Photo credits: (A and B) © Pest and Diseases Image Library; (C) © Mohammed El Damir. All from Bugwood.org).

grain before treatment. The webbing from Indian meal moth larvae also can jam and clog equipment in food processing facilities.

Mediterranean Flour Moth (*Ephestia kuehniella*)

The Mediterranean Flour Moth is less common than the Indian meal moths.

Adult moths are short-lived, do not feed and do not cause damage. Adults are pale gray with two distinct black zigzag lines on their wings (Figure 2). When the adult is at rest, the head and abdomen are slightly raised, making the wings look as if they slope downward. Adults fly at night in a very characteristic zigzag pattern.

The larvae of the Mediterranean flour moth are secondary feeders that prefer flour and meal. Female moths lay up to 675 white eggs on or near food. In 3 to 5 days, pinkish white larvae emerge. Larvae have reddish brown heads, a few small hairs and a few black spots on their bodies. They spin silken tubes within which they feed and mature. When fully developed, larvae are 1/2 to 2/3 inches long. Larvae are active crawlers. As they move, they spin webbed mats. Like Indian meal moths, these mats can clog processing equipment. They pupate near clean food, away from large amounts of infested material.

Figure 2: Mediterranean flour moth (Adult). Wings are about 1/2 inch long with a wingspan of a little less than an inch. (Photo © Mark Dreiling, Bugwood.org).

The Mediterranean flour moth also infests grain, nuts, seeds, and other stored foods.

Tobacco Moth (*Ephestia ellutela*)

The adult tobacco moth is light grayish brown with 2 light colored bands extending across each forewing. Its wingspan is about 5/8 inch. Black scales border the forewings. The hind wings are uniformly gray in color (Figure 3). Adult moths are short-lived and do not feed. They are attracted to lights at night and fly toward the top or roof of buildings at dawn and dusk.

Female moths lay up to 270 eggs singly or in small clusters on or near food. Larvae are pinkish to yellowish to off white. Larvae are about 1/2 inch long and migrate to sheltered areas to pupate. Found worldwide, the tobacco moth is a major pest of stored tobacco. It is a secondary feeder that attacks the highest priced tobacco grades. The larvae deposit silken webbing containing fecal pellets as they feed on cured tobacco leaves. When infestations are high, tobacco moths can consume entire leaves except for the midvein. Tobacco moth larvae also feed on a range of cereal, vegetable, and seed products.

Figure 3: Tobacco moth. The tobacco moth resembles a Mediterranean flour moth but is smaller. (Photo © John C. French Sr., Bugwood.org).

Beetles

There are a large number of beetle species that are pests of stored grain and food. Adult beetles of all species have 2 pairs of wings, but the first pair is thickened and hardened into wing covers. These wings meet to form a line down the middle of the back. Beetles undergo complete metamorphosis, with adults and larvae (also called grubs) having chewing mouthparts. Larvae have 3 pairs of legs. Some examples of beetle pests follow.

Red Flour Beetle (*Tribolium casteneum*)

The red flour beetle (Figure 4) is a serious pest in pasta and cereal manufacturing plants, flour mills, and bakeries where they infest cereal products, crackers, noodles, and other milled grain products. High populations of the beetle will discolor and alter the taste and baking quality of flour. Adults live 2 to 3 years. Red flour beetles are a serious pest of grain and have developed

Figure 4: Red flour beetle. The small size (1/7 of an inch) of these reddish brown beetles allows them to easily penetrate packaging and to hide in cracks in shelving. (Photo © Peggy Greb, USDA Agricultural Research Service, Bugwood.org).

resistance to fumigants in some parts of the U.S.

The design of a food production facility plays a role in infestations of red flour beetles and other pests. Dust can accumulate in food production facilities that have areas that are difficult to clean. These hard to clean areas are often where you will find beetles. Infestations have been traced to grain dust in light fixtures, candy machines, and voids in conveyer belts and other equipment. Flour and similar materials also drift and can be found on upper beams, ledges, ducts, and even inside equipment that appears to be sealed. Pheromone traps are used to monitor beetle activity. Constant monitoring and inspection is needed to detect problems early. In some instances, whole building fumigation may be needed to control a flour beetle infestation.

Drugstore (*Stegobium paniceum*) and Cigarette (*Lasioderma serricorne*) Beetles

The cigarette and drugstore beetles are two of the most common pests of stored products (Figure 5). Both species are found throughout the world. Both feed on a wide variety of stored food products but will also feed on non-food items such as leather, wool, potpourri, pinned insects, and other museum specimens. Most of the feeding damage from both species are caused by the larval stages. Adult cigarette beetles do not feed.

Mature larvae are up to 3/16 inch long, white, c-shaped, and almost cylindrical, with all body segments similar in size. Hairs covering the bodies of larval cigarette beetles are longer and more apparent than those on drugstore beetle larva. Larvae feed for 5 to 10 weeks before pupating.

Adults live up to six weeks. Adults of these beetles can penetrate most paper packaging (Figure 5-C). The adult drugstore beetle can even penetrate wood.

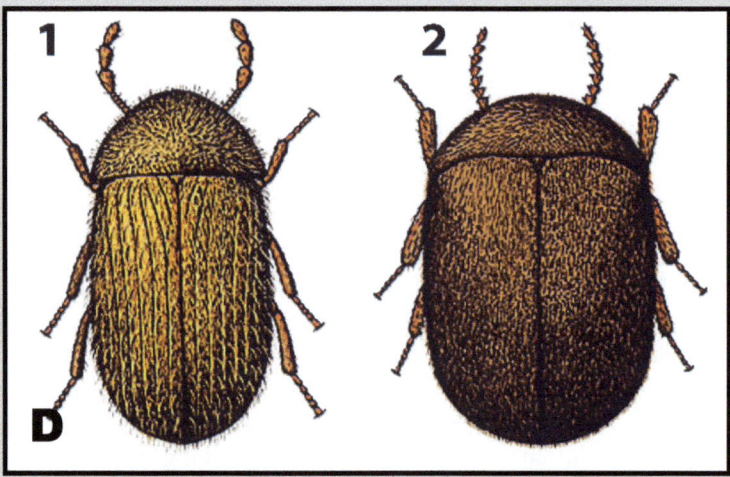

Figure 5: Drugstore and Cigarette Beetles. Drugstore beetles (A) and cigarette beetles (B) are small (1/16 to 1/8 of an inch long), cylindrical, reddish brown beetles that have their head tucked so that it cannot be seen from above. These beetles leave behind cylindrical holes in packaging when they emerge (C). These beetles have physical characteristics that distinguish them (D). The drugstore beetle has clubbed antennas with three broadened segments at the tips and wing covers that have distinct grooves (D-1). The cigarette beetle has sawlike or serrated antennas and wing covers that are smooth and lack distinct grooves (D-2). (Photo and illustration credits: (A) © Wood Products Insect Lab, USDA Forest Service; (B) © Brian Little, The University of Georgia; (C) © Gerald Holmes, California Polytechnic State University at San Luis Obispo; all courtesy of Bugwood.org. (D) © U.S. Department of Agriculture).

Granary Weevils (*Sitophilus granaries*)

Granary weevils are 1/8 of an inch long and dark brown with an elongated snout, a key feature of all weevils (Figure 6). Adult granary weevils cannot fly although other weevils that infest food can. Infestations are often seen in rice, whole grain corn, wheat, or sunflower seeds.

Sawtoothed Grain Beetles (*Oryzaephilus surinamensis*)
The sawtoothed grain beetle is another common stored product pest (Figure 7). It is a flat, dark brown beetle with 6 saw like teeth on each side of the thorax. The adult is 1/10 of an inch long.

Dermestid Beetles
"Dermestids" include over a dozen beetle species that are destructive (especially as larvae) to hides, skin, fur, wool, other animal related materials, and stored food. Examples of dermestids include the warehouse beetle (Figure 8), larder beetle, black carpet beetle, and furniture carpet beetle.

Adults are round and dark, and are usually about 1/4 of an inch long. The larder beetle is 1/2 of an inch long with a brown band around the middle.

Carpet beetles feed on grains, cereal, dried milk, chocolate, spices, hair, wool, feathers, silk, and dead insects (e.g., cluster flies). They can survive extended periods without food and are responsible for extensive damage to wool clothing. Larder beetles are often associated with dead birds and rodents, or high protein pet food. Carpet beetles can breed in horsehair insulation, bird nests, wasp nests, and grain based rodent baits. They are slower to breed than most stored product pests.

Figure 6: Granary weevils. Granary weevils and related species can only breed on whole grain or seeds. (Photo credits: (top) © Clemson University - USDA Cooperative Extension Slide Series; (bottom) © Pest and Diseases Image Library; both courtesy of Bugwood.org).

Figure 7: Sawtoothed grain beetles. These beetles can produce up to 6 generations per year. Populations are most successful when the relative humidity is 70% or higher. (Photo © Emilie Bess, USDA APHIS PPQ, Bugwood.org).

Comparing Beetle and Moth Larvae
Adult beetles and moths can easily be differentiated by their appearance. Beetles have a hard shell while moths have two pair of fuzzy wings. Their larvae, on the other hand, can be hard to distinguish. A key difference between moth and beetle larvae is the presence of distinct non-jointed, fleshy "prolegs" further down the abdomen in caterpillars (moths). Moth larvae have three pairs of true, jointed legs just behind their head and varying number of prolegs. Some beetle larvae also have three pairs of jointed true legs behind the head, but others do not. Beetle larvae, however, do not have any prolegs (Figure 9).

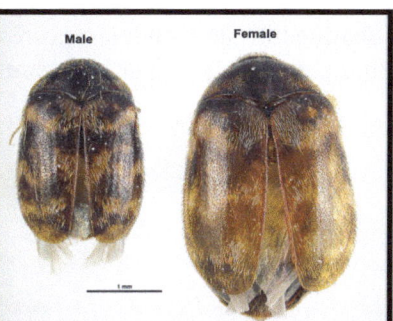

Figure 8: Dermestid beetles. Warehouse beetles (A and B) have 1/8 inch long oval bodies with a brown and yellowish pattern on the wing covers. Dermestid beetle larvae (C) are typically golden to dark brown, hairy, and have an elongated shape tapered on both ends. (Photo credits: (left and middle) © Pest and Diseases Image Library; (right) © Whitney Cranshaw, Colorado State University. All from Bugwood.org)

Grain Mites

Several types of grain mites will breed indoors or on food products if conditions are damp. These mites require a relative humidity above 80% to survive. Drying out a room or building will often manage the mite problem. However, drying may not be feasible for infestations in cheese plants or high moisture animal feed. In these situations, fumigation might be the only practical solution to control grain mites.

Livestock may refuse to eat grain that has been heavily infested with grain mites, even after the mites have been killed. In this instance, disposing of infested grain might be the only option.

Control of Grain and Commodity Pests

Both primary and secondary feeders can cause significant financial losses to growers of raw agricultural products and owners of food processing facilities. In some situations, fumigation may be one of several control methods to consider, but in others, it might be the only viable option.

Prevention is key to controlling grain and commodity pests. Regularly clean in and around product storage areas. Practice good sanitation by removing spilled grain, grain dust, and other stored product inside and outside the storage area. Be sure to clean corners, floors, and walls. Bits of product remaining can harbor insects that can move into new products stored in the same area. Apply non-fumigant pesticides in areas that are hard to reach. Routine cleaning, along with preventive spot fumigation, can reduce or eliminate the need for large scale fumigation.

Figure 9: While different species of both moth and beetle larvae vary in appearance, two that can look similar are the Indian meal moth caterpillar (top) and the red flour beetle larva (bottom). The main distinguishing feature is that while moth caterpillars have prolegs on their abdomen, beetle larvae do not. (Illustration credit: © UW-Madison Pesticide Applicator Training Program. Photos for illustration (moth larva) © Todd Gilligan, CSU, Bugwood.org; (beetle larva) © John Obermeyer, Purdue Extension).

Follow good storage practices. Monitor for insect infestations and temperature increases in stored products and grain. Aerate the storage facility to maintain low moisture levels in the grain. Mix and level the grain so fines and other grain debris are evenly distributed throughout the grain. If an infestation occurs, first attempt to find and destroy all infested products, then treat the location where they were stored. If the infestation is severe, you might need to fumigate. Practice good housekeeping to prevent reinfestation.

Assess the effectiveness of a fumigation by comparing insect activity or presence before and after the treatment. A convenient and fairly accurate method is to use pheromone traps, or insect bioassays. In insect bioassays, one or more stages of lab reared live insects are placed in a ventilated vial or other container and placed within the structure or grain prior to treatment. After the treatment, the bioassays are examined to confirm that the fumigant applied killed the insects. Sealing bins and other structures as tightly as possible is essential in all fumigations, but especially in commodity fumigations where insecticide resistance has been increasingly observed.

Chapter 8.3 Review Questions

Correct answers are given on page 175.

1. Which of the following is a secondary feeder?
    a. cigarette beetle
    b. maize weevil
    c. lesser grain borer

2. An important practice in the control of stored commodity pests is _____.
    a. maintaining a high humidity in all storage areas
    b. practicing good sanitation in and out of storage areas
    c. applying a broad spectrum insecticide to storage areas on a quarterly basis

CHAPTER 8.4

Miscellaneous Pests

LEARNING OBJECTIVES

☑ Discuss the importance of proper pest identification of rodents and microorganisms.
☑ Discuss management options for rodents and microorganisms.

Terms to Know
The following are important terms to know from this chapter. They are explained and *italicized* in the text and defined in the glossary at the end of this manual.

Brown Rot	White Rot
Soft Rot	Lignin

Miscellaneous Pests

Some pests controlled by non-soil fumigation methods do not neatly fit into broad categories used in this study manual. These include rodents and microorganisms (i.e., bacteria, viruses, and fungi). Rodents can be major problems in farm fields, in stored commodities, and other infrastructures (e.g., levees, ditches). Microorganisms (e.g., fungi and bacteria) can damage wood, and contaminate food and other products such as wine corks and barrels (as discussed in part in previous chapters).

This chapter will cover some of the common miscellaneous pests you might encounter in your work. It is not an exhaustive list, nor is it meant to be a definitive pest identification manual. Accurate identification of these miscellaneous pest species is essential in pest management and will help you to determine the best control or management method (i.e., fumigation), the pesticide active ingredient(s) to use, the correct dosage, etc. If necessary, consult an expert (i.e., biologist, extension worker) to help with pest identification. Always read and follow the pesticide label for use information.

In California, the Structural Pest Control Board issues the license for the use of fumigants, such as sulfuryl fluoride, to control structural pests. This study manual does not cover these uses. Applicators fumigating stored commodities or rodent burrows must hold the DPR Non-Soil Fumigation category. Note, if you work for a government sponsored program for the management and control of pests having public health importance using non-fumigant pesticides, DPR's Health Related (Category K) may also be needed. In addition, if you are an employee of a government mosquito or vector control agency, consult the Department of Public Health to ensure you hold the proper licensure to control vertebrates. However, certification as a Vector Control Technician alone does not permit the use of fumigants to control rodent vectors.

Several resources are available to help accurately identify some of the pests covered in this chapter. The University of California Statewide IPM Program's Pest Notes Library or the University of Nebraska's Prevention and Control of Wildlife Damage provide recommendations on pest control options for various pest species.

Rodents

Rodents cause economic damage and may also be a public health concern. Rodents, such as ground squirrels, can compete with cattle for rangeland forage and are common pests of alfalfa, nut, and citrus crops. Rodent also cause damage by their burrowing and gnawing. Gnawed electrical lines can cause fires, power outages, or telephone service interruptions. Burrows can cause physical injury to livestock and humans. Rodents can consume and contaminate large amounts of stored products. One rat can consume about 50 pounds of grain a year but can contaminate much more.

Commensal rodents can also vector diseases. Sylvatic plague and murine typhus can be transmitted by infected fleas that jump hosts from rodents to humans. Other diseases can be spread by other arthropod vectors (e.g., West Nile Fever, Malaria), fecal contamination (e.g., Salmonellosis, E. Coli), or bites.

The most serious rodent pests causing damage to stored products are the Norway rat, roof rat, and the house mouse. Rodent control can be achieved through means other than the use of pesticides. Even when pesticides are used, fumigants are not common tools. However, fumigants are used in some circumstances.

Rats and Mice

Two rat species (the Norway rat, *Rattus norvegicus*, and the Roof rat, *Rattus rattus*) and the house mouse (*Mus musculus*) are briefly discussed in this chapter.

Norway Rat (*Rattus norvegicus*)

Sometimes referred to as the brown rat or sewer rat, the Norway rat (*Rattus norvegicus*) creates open system burrows around buildings, under debris or piles of wood, and in fields with areas of high moisture. These burrows contain nests often lined with scraps of paper, insulation, and other shredded materials. These rats weigh approximately 12 to 16 ounces and are about 16 inches in length, including their tail. You can identify Norway rats by their blunt snouts, small beady eyes, small ears, shaggy fur, and scale like tail that is shorter than the length of the body and head combined. Do not confuse the Norway rat with the roof rat, *Rattus rattus*. The Norway rat is larger and more aggressive than the roof rat and unlike roof rats, Norway rats have tails that are only as long as their body.

The breeding season of Norway rats peak in the spring and the fall, although they are capable of breeding year round. Female Norway rats reach sexual maturity between 2 and 3 months of age and have 4 to 6 litters per year with 8 to 12 offspring per litter. These rodents have an average lifespan of about one year. They are social animals and prefer to live in colonies. Norway rat families can have several connected burrows with one main hole that is used as an entrance and an exit, as well as a few additional emergency exit holes.

You can recognize Norway rat infestations by the presence of shiny black capsule shaped scat with blunt ends that measure on average 3/4 inch in length. Norway rats' hind foot tracks measure about 3/4 to 1 inch and they leave a trail from dragging their tail in between their feet. Their burrows can be identified by their smooth walls with entrances that are 2 to 4 inches in diameter. Norway rats cause damage to asparagus, avocados, citrus, corn, olives, melons, squash, nuts, and grain and cereal crops. Most damage inflicted by Norway rats are to post-harvest or food storage. One rat eats approximately 20 to 40 pounds of feed per year. Besides eating stored produce and grains, Norway rats contaminate stored food which will usually have to be destroyed. They can also chew on electrical lines and hoses to damage engines.

Roof Rat (*Rattus rattus*)

Roof rats (*Rattus rattus*) are sometimes referred to as black rats or ship rats. They are smaller than Norway rats and their tails are longer in comparison to the length of their head and body. These rats average 12 to 16 inches in length from the nose to the tip of the tail and weigh 5 to 9 ounces. You can identify these rats by their brownish black coloration, long and thin body, pointy nose, large eyes, protruding ears, and black, scaly tail. Roof rats breed year round provided conditions are favorable, and breeding peaks in the spring and fall months. Female roof rats reach sexual maturity at 3 to 5 months of age and can produce five litters per year consisting of 5 to 8 offspring per litter. The roof rat's lifespan is approximately one year. Roof rats do not burrow often, as they are exceptional climbers and prefer to nest in areas up high, such as trees or attics.

Signs of roof rat infestation include long cylindrical droppings (measuring about ½ inch with pointed ends), gnaw marks, and greasy rub marks. Roof rats are exceptional climbers and can cause damage to commodities such as citrus, avocados, pomegranates, and nut crops. Aside from feeding on the fruit and rendering it unmarketable, roof rats also cause damage by chewing on tree limbs, causing branches to die. Roof rats, like the Norway rat, can also contaminate stored food (e.g., grains). Roof rats can also chew on, and damage, electrical wires and irrigation lines.

House Mouse (*Mus musculus*)

House mice are small, have slim bodies, and are about 5½ to 7½ inches long. They have a pointed muzzle, large eyes, and large ears. They tend to be colored gray or brown. House mice prefer cereal grain, but they will eat almost anything. They can live in areas with no water present because they can get their water needs from the food they eat. House mice will live almost anywhere, from cabinets, to empty spaces between walls. They also inhabit barns, stables, grasslands, patches of vegetation, or garbage piles. Mice can cause considerable damage to homes by chewing through walls, nesting in electrical boxes, and chewing through wires.

Differences Between Rats and Mice

Table 8.4-1 summarizes some of the differences between these species. Knowing the physical differences between these rodent species as well as their traits and habits is important in determining how best to manage them. Size is an easy factor to tell rats and mice apart. However, an adult mouse and a juvenile rat can be about the same size and could be confused for each other based on size alone. Young rats have a relatively large head and large feet for their body size compared to mice. In addition, the head of a juvenile rat is short and stubby while an adult mouse has a more triangular shaped head. The juvenile rat's muzzle is relatively large, blunt, and wide compared to the adult mouse's narrow, sharp muzzle. Also, the juvenile rat's tail is relatively thick compared to the adult mouse's relatively thin tail.

Ground Squirrels

The California ground squirrel (*Otospermophilus beecheyi*) is the most common ground squirrel species considered to be a pest in California. This squirrel can be found in nearly all regions of the state. California ground squirrels prefer open grass lands but are also commonly found in grain fields, irrigated pastures, meadows, open fields, and around home and residential areas.

California ground squirrels are typically 14 to 20 inches in length, including their tail. Adults weigh between 21 and 30 ounces with the males being slightly larger than the females. Their fur is brown with white and gray markings on their backs, and they have a white ring around each eye. Breeding season for California ground squirrels is dependent on climate and location. Those that live in colder areas at higher altitudes will hibernate longer, thus postponing the breeding season. In areas where the weather is warmer, breeding season can last from January to July. Female California ground squirrels reach sexual maturity at about one year of age and produce one litter per year with an average of 5 to 8 young per litter.

They are social animals that form colonies and live in open system burrows. Ground squirrel burrow openings are about 4 to 6 inches in diameter, with multiple openings connecting a series of tunnels. This species can cause damage to irrigation systems, crops, and structures.

Pocket Gophers

Pocket gophers (family *Geomyidae*) are named for their fur lined cheek pouches. These gophers are adapted for foraging and burrowing. They have clawed front paws for digging, small eyes and ears, and can range in size from 6 to 10 inches. Pocket gophers can be found throughout California commonly in woodland, chaparral, scrubland, and agricultural terrains.

Pocket gophers do not hibernate, although you will not often find them outside of their burrows unless they are foraging, pushing dirt out of their burrow, or relocating. Pocket gopher burrows are closed system burrows. They can be easily identified by the crescent shaped mound of dirt surrounding the plugged-up entrance to their burrow. Their burrow openings range from 2½ to 3½ inches in diameter and the burrow itself can cover an area up to 2,000 square feet. Dry, loose soil is typically a poor pocket gopher habitat as the burrow systems can collapse easier compared to moist soil. Pocket gophers also prefer moist soil because they normally feed on plant roots they come across while creating their burrows.

These gophers are active year-round, at all hours of the day. These rodents reach sexual maturity in about one year and produce 1 to 3 liters per year with a litter consisting of 5 to 6 offspring. Pocket gophers are typically solitary and are only found with other pocket gophers during breeding season and when raising young. Pocket gophers can cause damage to structures, irrigation pipes, and ditches from their burrowing and girdling activities.

Rodent Management With Fumigants

In structures, effective rodent management requires a multi-method approach involving sanitation, rodent exclusion (e.g., closing holes, mesh or foam around pipes, etc.), traps, and rodenticide baits. However, in a large space, such as a grain silo, fumigants could be the most cost effective way to reduce rodent numbers. Sulfuryl fluoride and aluminum phosphide are labeled for rodent control in those structures.

Retrieve as many rodent carcasses as possible after a fumigation. Carcasses that are not removed are unsanitary, will decompose and cause odors, and may be a breeding ground for insects. Burrowing rodents, such as pocket gophers and ground squirrels can be controlled with burrow fumigation using aluminum phosphide or carbon dioxide.

Always be aware that the pest control method you use, including fumigation, may affect non-target animals, including legally protected species. For example, the blunt-nosed leopard lizard (*Gambelia silus*) uses small mammal burrows for shelter. The lizard is a federally endangered species found in the following counties: Merced, Madera, Fresno, San Benito, Kings, Tulare, Kern, San Luis Obispo, Santa Barbara, and Ventura counties. Fumigants labeled for use in outdoor burrows may have a statement requiring applicators to obtain endangered species bulletins for the area where the treatment will occur.

See Chapter 7.3 for additional information on Burrow Fumigation for Vertebrate Pests.

Figure 1: Rats and Mice. Rodents are the most successful mammals in terms of total numbers and variety of species. "Commensal rodents" have become so adapted to the human dominated environment and benefit from being around humans." The Norway rat (*Rattus norvegicus*) (A) and the house mouse (*Mus musculus*) (B) are the most common commensal rodents found in the U.S., although the roof rat (*Rattus rattus*) (C) is quickly becoming a more prevalent problem. (D) The deer mouse (*Peromyscus maniculatus*) and the similar looking white-footed mouse (*P. leucopus*: not pictured) are two native mice that are confused with house mice and occasionally cause structural damage. (Photo credits: (A) © Matt Frye, New York State IPM Program, Cornell University; (B) © David Cappaert, Bugwood.org; (C) © Niamh Quinn); (D) © Joellen Lampman, New York State IPM Program, Cornell University).

Microorganisms

Microorganisms (or "microbes" for short) are tiny forms of life usually not visible to the naked eye such as fungi, bacteria, and viruses. Our world is full of microbes and most do no harm to people or products. Some are beneficial (we wouldn't have wine, beer, cheese, bread and much more without certain fungi and bacteria). Some microbes, however, are harmful and can damage or contaminate commodities.

We usually manage microbes in ways that either don't involve pesticides, but when we use pesticides, they are not usually fumigants. However, there are some situations where fumigants are used to control certain microbes including fumigation of utility poles to control fungi and sanitizing wine barrels and wine bottle corks to reduce the number of microorganisms that could disrupt the aging process, impart undesirable flavors to the wine, or cause spoilage. These uses were discussed in Chapters 7.8 (Sanitizing Wooden Wine Barrels and Wine Bottle Corks) and Chapter 7.9 (Remedial Wood Protection).

Fumigants used to control miscellaneous microorganism pests include sulfur dioxide (SO_2), MITC, metam sodium, and dazomet.

Fungi

Fungi are neither plant or animal and belong to their own Kingdom. Fungi are diverse with millions of species. Examples of fungi include yeasts, molds, mildews, rusts and mushrooms. Fungi break down organic matter (either living or dead) to feed themselves. Most of the mass of a fungus is composed of long, threadlike filaments called hyphae. The hyphae are invisible to the naked eye and absorb water and nutrients needed for growth and reproduction of the fungus. The fruiting bodies (what we commonly call mushrooms) or masses of hyphae (collectively called "mycelium") are the parts of fungi you can see without a microscope and are more likely to notice.

Some fungi are parasitic to plants (i.e., mildews, scabs, or canker) and can lead to economic losses in crops. A small number of fungi can cause diseases in animals, including humans (i.e., Coccidioidomycosis, Histoplasmosis). Other fungal species can cause decay and mildew in wood, fabrics, and paper products.

Some fungi digest wood, either living or dead. In nature this helps recycle trees and frees up nutrients for growing plants to use. However, when the affected wood is part of a structure (e.g., utility pole, wooden food storage container, wooden wine barrel) that destruction interferes with the wood's use.

Fungi that grow on wood have two basic requirements: temperature, and moisture. Temperature: The most favorable temperature for fungi to grow on wood is between 70° to 85°F although fungal growth can occur between 50° to 90°F. Temperatures below 35°F and above 100°F are generally safe from decay. Moisture: Fungi require a wood moisture content of about 30%. Fungi will not attack dry wood that has a moisture content of less than 19%.

Table 8.4-1. Adult Rodent Characteristics			
Characteristic	**Norway Rat**	**Roof Rat**	**House Mouse**
Body weight	7 - 18 ounces	5 - 10 ounces	0.5 - 1 ounce
Total length	12 - 18 inches	13 -18 inches	5 - 8 inches
Tail length	5 - 8 inches	7 - 10 inches	2 - 4 inches
Tail	Shorter than body, paler below, carried stiffly behind the animal, coarse scales.	Longer than their heads and bodies combined.	Equal to or a little longer than body-head length, paler below.
Ears	Close set, short, thick with short fine hairs, small relative to head. Don't reach eyes if folded over.	Large, nearly naked. Long enough to reach eyes if folded over.	Large relative to head with some hairs.
Hind foot length	~ 1.7 inches	~ 1.3 inches	Under ¾ inch
Number teats on female	12	10	10
Color of belly fur	White with gray underfur.	Uniform: all white, all buff, or all gray.	Variable, but lighter in color than body fur.
Overall appearance	Large, robust body. Small eyes.	Sleek, agile body. Large eyes.	Fine brownish gray to gray.
Droppings	Up to 3/4 inch	Up to 1/2 inch	Up to 1/4 inch

Fungi that sustain on or in wood are categorized into two groups: wood destroying fungi (also known as decay fungi), and wood staining fungi.

Wood destroying fungi can cause decay in both sapwood and heartwood. However, sapwood of all tree varieties degrade in warm, moist soil. This type of fungi can further be categorized as *brown rot*, *white rot*, and *soft rot* fungi. Brown rot fungi breaks down cellulose in wood, which leaves behind a brown residue. Brown rot fungi are the most common cause of decay of softwood tree species. White Rot Fungi breaks down cellulose and *lignin* and has a bleaching effect. White rot fungi are seen more in hardwood tree species. Soft rot fungi attacks wood that has absorbed high amounts of water that causes weakening and softening of wood.

Figure 2: Wood destroying fungi. Note the shelflike appearance.

Wood destroying fungi can appear on wood surfaces or grow in the surrounding environment of the wood. The color of wood destroying fungi may range from white to dark brown, light brown, or bright yellow. The fungi may appear flattened and/or shelflike (Figure 2). Wood destroying fungi can stop growing with changes in temperature and moisture.

Wood-staining fungi cause discoloration in wood and includes sap-staining and mold fungi. Sap-staining fungi causes sapwood, especially in softwood species, to show differences in wood color. This fungi is usually transported to trees via beetles. Mold fungi has a fuzzy-like texture and is noticeable by its color; green, yellow, brown, or black. Mold fungi can form in humid or wet weather. Removing the moisture source can help control this type of fungi.

How Fungi Spread. Spores of fungi spread by air, water, insects, and any practice that moves the infected host from one location to another. A fungal organism requires certain environmental conditions to begin to grow and infect objects. These conditions often include high humidity or the presence of water and warm temperatures.

Bacteria

Bacteria are microscopic, one celled organisms that reproduce by dividing to become two identical cells, a process known as fission. Some bacteria produce spores that make them more difficult to control, as spores make the microbe extremely resistant to heat, pesticides, and drying.

Bacteria can be characterized by their shape and/or the way they absorb special stains. Bacteria assume one of five shapes: spherical (cocci), rod shaped (bacilli), comma shaped (vibrio), spiral shaped (spirilla) or corkscrew (spirochete). Gram staining is used to differentiate between "gram positive" bacteria that take up the stain and "gram-negative" bacteria that do not.

Some bacteria can cause diseases and infection in plants, animals, and humans. Other bacterial species can produce enzymes (a protein that speeds up a chemical reaction) that can damage non-living materials such as adhesive and plastics or contaminate food and food processing equipment.

How Bacteria Spread. Bacteria can spread through air, water, or soil, and via insects, animals, humans, or objects (i.e., equipment, clothing, shoes) that move or are transported from one location to another. Unlike viruses, bacteria can survive in the environment for relatively longer periods.

Viruses

Viruses are extremely small organisms that can only be seen through an electron microscope. Viruses reproduce inside living cells. Controversy exists on whether viruses are living organisms since they cannot grow, are incapable of producing their own energy, and require a host plant or animal cell to survive. Most viruses can only survive for short periods outside a host plant or animal cell. When viruses invade living cells, they use their own genetic information to alter these cells, so the cells produce more viruses rather than their usual proteins or nucleic acid.

Viruses alter chemical activity within host cells and these changes often cause disease.

How Viruses Spread. Viruses spread through the air, water, bodily fluids, tissues, and on contaminated fomites (inanimate objects such as equipment, clothing, footwear, etc. that transmit and can transfer disease to a new and susceptible host).

Controlling Microorganisms

Methods for controlling microorganisms include chemical (antimicrobial pesticides) and non-chemical approaches. Antimicrobial pesticides either destroy the microorganisms or provide barriers that exclude them from an area. Sometimes antimicrobials inhibit the growth of microorganisms. Non-chemical methods include exclusion and sanitation and may also involve modifying the environment to regulate temperature and moisture. Other methods, such as irradiation, also destroy certain microorganisms.

Exclusion

Physical barriers such as walls or partitions exclude certain microorganisms from some areas. Thoroughly cleaning objects before moving them to other locations helps to prevent spreading microorganisms. Positive air pressure in a confined area also keeps airborne microorganisms out. High efficiency filters in air conditioning systems trap microorganisms and prevent their spread to other areas within a building. Special water filters remove microorganisms from water pipes.

Sanitation

The first step in sanitation involves cleaning surfaces to remove foreign materials which can otherwise interfere with effective pest control by shielding the microorganisms. Cleaning usually requires washing or scrubbing with soap and water or other cleaning material. Following cleaning, treating the surfaces with a disinfectant material or heat greatly reduces the population of microbial organisms.

Modifying the Environment

Maintaining temperatures at sufficiently high or low ranges suppresses or destroys microorganisms. This is why refrigeration or heat protects foods. Some microorganisms cannot survive unless free water is available; therefore, keeping an area dry will prevent such organisms from reproducing. Submerging objects in boiling water for several minutes kills most microorganisms, although some bacterial spores are highly resistant to boiling water. Saturated steam under pressure, known as autoclaving, is a reliable method used in hospitals for killing microorganisms. Heating objects with a flame or applying dry heat to them is also effective in destroying most microorganisms.

Irradiation

Irradiation with ultraviolet light, x-rays, or gamma rays effectively destroys microorganisms, but requires highly specialized and potentially hazardous equipment that trained personnel must operate. Irradiation is confined to small areas for safety purposes.

Chemical Control

Applying antimicrobial pesticides is usually a successful and inexpensive way to control many types of microorganisms. Additionally, combining pesticide applications with certain non-chemical methods typically results in control that is more effective. For example, to control molds, steps can be taken to reduce or eliminate moisture to supplement applications of antimicrobials.

Due to their broad pest control properties, some fumigants may be used as antimicrobials. For example, as discussed in Chapter 7.9 MITC and MITC-generating pesticides (metam sodium and dazomet) are sometimes used to treat wood, such as utility poles, pilings, and bridge timbers, to control wood destroying fungi and prevent decay. The wine industry uses sulfur dioxide (SO_2) as a fumigant to sterilize wine barrels and corks, as discussed in Chapter 7.8. In many cases, the combined use of chemical and non-chemical methods is more effective because single methods of control rarely eliminate all of the organisms. Because of health hazards caused by certain microorganisms, total eradication or sterilization is the primary goal of a management approach, although this may be difficult to achieve.

Chapter 8.4 Review Questions

Correct answers are given on page 175.

1. Which of the following rodent pests causes serious damage to stored products?
    a. Norway rat
    b. Western rock squirrel
    c. Pilliga mouse

2. Pocket gopher burrows are _____.
    a. open systems
    b. above ground systems
    c. closed systems

3. Which of the following pesticides is labeled for rodent control in a structure, such as a grain silo?
    a. MITC
    b. metam sodium
    c. sulfuryl fluoride

Correct Answers to Chapter Review Questions

Chapter 1
1. A
2. A
3. C
4. B
5. A

Chapter 2
1. C
2. B
3. C
4. A

Chapter 3
1. B
2. B
3. B
4. C
5. B

Chapter 4
1. C
2. A
3. A
4. A
5. B

Chapter 5
1. B
2. C
3. A
4. A
5. A

Chapter 6
1. B
2. A
3. B
4. A
5. A

Chapter 7.1
1. C
2. A
3. A
4. B
5. C

Chapter 7.2
1. B
2. C
3. A

Chapter 7.3
1. C
2. A
3. A

Chapter 7.4
1. C
2. B
3. B

Chapter 7.5
1. A
2. C

Chapter 7.6
1. B
2. B

Chapter 7.7
1. B
2. C
3. B
4. A

Chapter 7.8
1. B
2. A

Chapter 7.9
1. A
2. A
3. C

Chapter 8.1
1. A
2. C
3. A

Chapter 8.2
1. C
2. A
3. B
4. A

Chapter 8.3
1. A
2. B

Chapter 8.4
1. A
2. C
3. C

Glossary

Absorb, Absorbed	(1) The entrance or taking up of a pesticide into a body through the skin, eyes or mouth. (2) In the case of a fumigant, when the molecules penetrate into a material (commodity, soil, wood, etc.).
Active ingredient	The chemical or chemicals in a pesticide formulation that are biologically active and are capable, in themselves, of preventing, destroying, repelling, or mitigating insects, fungi, rodents, weeds, or other pests. The remainder of the product consists of one or more inert ingredients (such as water, solvents, emulsifiers, surfactants, clay and propellants), which are there for reasons other than pesticidal activity.
Actual flow	The actual volume of water that enters the treatment plant on a given day.
Acute effects	Illness or injury that may appear immediately after exposure to a pesticide (usually within 24 hours).
Acute exposure	Exposure to a single dose of a pesticide. A one-time event.
Acute toxicity	A measure of the capacity of a pesticide to cause injury as a result of a single exposure.
Adsorption	The process by which a substance sticks to the surface of something else (e.g., when a fumigant binds to the surface of a commodity).
Aeration	In relation to fumigation, it is the process of adding air or allowing air into the space fumigated (or the container that held the fumigant) to allow the fumigant to dissipate to safe levels.
Aeration buffer zone	An area that extends from the point of fumigant emission from the treatment area (e.g. exhaust stack or building edge) to a specified distance where access is limited.
Aeration period	The period of time starting at the initiation of aeration and ending when the concentration of the fumigant, as measured according to label instructions, reaches the level noted on the label and the minimum aeration time has elapsed.
Aerosol	A suspension of fine solid or liquid particles in air. NOT the same thing as vapor or gas.
Agitation	The process of stirring a pesticide solution to keep the components in suspension.
Agricultural commodity	A legal definition from Title 3, California Code of Regulations section 6000. It means an unprocessed product of farms, ranches, nurseries, and forests, except live-stock, poultry, and fish. Agricultural commodities include: • Fruits and vegetables; • Grains, such as wheat, barley, oats, rye, triticale, rice, corn, and sorghum; • Legumes, such as field beans and peas; • Animal feed and forage crops; • Rangeland and pasture; • Seed crops; • Fiber crops, such as cotton and fax; • Oil crops, such as safflower; • Sunflower, corn, and cottonseed; • Trees grown for lumber and wood products; • Nursery stock grown commercially; • Christmas trees; • Ornamentals and cut flowers; and • Turf grown commercially for sod

Agricultural use	A legal definition from California Food and Agricultural Code section 11408. This means the use of any pesticide or method or device for the control of plant or animal pests, or any other pests, or the use of any pesticide for the regulation of plant growth or defoliation of plants. This term excludes the sale or use of pesticides in properly labeled packages or containers that are intended for any of the following: • Home use * • Use by structural * pest control operators • Industrial use* • Institutional use* • Use under a veterinarian's prescription (animal pests only), and • Use by a vector control district or agency under a cooperative agreement with the California Department of Public Health "Agricultural use" includes, but is not limited to, commercial production of animals or plants, forests, parks, golf courses, cemeteries, roadsides, rights-of-way, and nurseries. * Home, structural, industrial, and institutional uses are defined in Title 3, California Code of Regulations section 6000.
Air monitoring	The use of sensitive gas monitoring devices during fumigation to accurately gauge the dosage of the fumigant in the treated area and/or to detect leaks from the application site.
Ambient air analyzer	A type of monitor that uses infrared light to detect and measure gas fumigant concentrations. Also called "IR analyzer."
Antimicrobial	A substance that is intended to disinfect, sanitize, reduce, or mitigate growth or development of microbial organisms.
Applicator's Manual	Most, if not all fumigant labels direct the applicator to use the product's Applicator's Manual. That manual is much longer and contains more detailed instructions on use of the product than the label attached to the product. Just as for the label, the applicator's manual is a legally-binding document and the instructions must be followed explicitly.
Atmosphere-supplying respirator	A device that draws air from outside a fumigation area or uses canisters of pressurized air to supply a worker with breathable air. The latter is also called a self-contained breathing apparatus (SCBA).
Atmospheric chamber	A structure used to conduct a fumigation that is under normal (i.e., ambient) air pressure.
Autoclaving	A process that uses high pressure saturated steam for several minutes to sterilize objects.
Bacterium	A single-celled microorganism that lives in soil, water, organic matter, or the bodies of plants and animals. Some bacteria cause plant or animal diseases (plural: bacteria).
Boiling Point	The temperature at which the vapor pressure of a liquid equals the pressure surrounding the liquid and the liquid changes into a vapor. Or, simply, the temperature at which a liquid becomes a gas.
Brown rot	A type of wood destroying fungi that remove cellulose, leaving wood darkened and fractured.
Buffer zone	A restricted-access area established around the perimeter of an area being fumigated. The certified applicator (and authorized fumigation handlers under his/her direct supervision) must prohibit entry into the buffer zone by any other person. Note that EPA has adopted the terms "treatment buffer zone" and "aeration buffer zone" to designate the zones in effect during the application and exposure periods (the "treatment" time) and during aeration. These will differ in scope and may differ in location.
Building sewer line	The portion of a sewer system which lies between the building foundation and a main sewer line; also called lateral sewer line.
Burrow fumigation	A type of fumigation used to control certain vertebrates in outside burrows.
Calibration	The process of adjusting application equipment so that pesticides are applied at a known prescribed rate. Calibration is also performed on air monitoring devices to ensure accurate readings of fumigant concentration.
California Department of Pesticide Regulation (DPR)	The state lead agency responsible for regulating the use of pesticides in California.
Canister	A device used with a respirator that contains components that absorb specific gases. Each canister is color coded with stripes to indicate limitations and approved uses.
Canister respirator	A respirator that uses canisters to remove toxic fumes from air.

Cellulose	The carbohydrate that is the principal constituent of wood and component of wood cells.
Certified applicator	A person who has demonstrated, through an examination process, the ability to safely handle and apply highly hazardous restricted pesticides.
Certified applicator-in-charge	A certified applicator who has supervisory authority over the fumigant application. Note that EPA uses the term "site supervisor."
Chamber fumigation	The use of a well-sealed structure to conduct a fumigation. Some chambers are specially built for fumigation, while others are modified rooms or buildings.
Check	A lengthwise separation of the wood that usually extends across the rings of annual growth and commonly results from stresses in wood that occur during seasoning.
Chemical name	The scientific name for a chemical substance. For example: *sodium methyldithiocarbamate* is the chemical name for metam sodium.
Chemical reactivity	The tendency of a substance to undergo chemical reaction, either by itself or with other materials. One common example of a chemical reaction is when iron combines with oxygen to make iron oxide, or rust.
Chemical-resistant	A material that allows no measurable movement of the pesticide through it during use.
CHEMTREC	The chemical transportation emergency center. This organization operates a 24 hour information hot-line for pesticide spills, fires, and accidents. 1-800-424-9300.
Chronic	Pesticide-related illness or disease that may extend over months, years, or a lifetime.
Chronic toxicity	The potential for long-term health effects as a result of exposure to a particular pesticide.
Clean out	A capped opening into a building sewer lateral line providing access for cleaning equipment.
Collector sewer	A sewer, typically small diameter, which collects wastewater flows from buildings and transports those flows to an interceptor sewer.
Combined sewer	A sewer which is designed to carry both sanitary flows and storm water, either all or part of the time.
Compatibility	The ability of two pesticides or substances to mix without reducing the effectiveness or usefulness of either substance.
Concentration	The amount of a substance in a given weight or volume.
Contact herbicide	An herbicide that kills primarily by its contact with plant tissues rather than by being translocated to other parts of the plant. For sewer line root control, metam sodium is a contact herbicide.
Corrosive	Causing damage by chemical action such as when a substance oxidizes (e.g., rusts) a metal surface, or corrodes like an acid. Some fumigants are especially corrosive to certain metals.
County Agricultural Commissioner (CAC)	The official in each county in California who has the responsibility for enforcing the state and federal pesticide regulations and issuing permits for restricted-use pesticides. County agricultural commissioners and their staff frequently inspect pesticide applications and application sites and conduct investigations into complaints of pesticide misuse.
CT (Concentration x Time) Concept	The dosage required to kill the target pest(s) that is accumulated over a period of time and measured in ounce-hours or gram-hours.
Decay	The decomposition of a wood substance by fungi. In advanced decay, destruction is recognized readily because the wood has become punky, soft and spongy, stringy, ringshaked, pitted, or crumbly. Obvious discoloration or bleaching of the rotted wood often is apparent.
Decomposition/degradation	The process by which a chemical substance is broken down into simpler substances. This process can take place through chemical, biological, or physical means.
Dermal	Pertaining to the skin. One of the major ways pesticides can enter the body to possibly cause poisoning.
Design flow	The amount of wastewater that the treatment plant is designed to handle daily.
Desorb / Desorption	The process of a substance releasing from or through a surface. The process is the opposite of sorption.
Detector tubes	A glass tube that shows a color change in the presence of a specific gas such as methyl bromide. Tubes are specific for different fumigants. Also called "colorimetric tubes."

Detectors	A generic term for any of various monitoring tools used to measure the presence of a substance (e.g., concentration of phosphine in air).
Diffuse / Diffusion	Diffuse: to cause a gas or liquid to spread through or into a surrounding substance by mixing with it. Diffusion: a process by which there is a net flow of matter from a region of high concentration to a region of low concentration resulting from random motion of molecules.
Disease	A condition, caused by biotic or abiotic factors, that impairs some or all of the normal functions of a living organism.
Dosage	The addition of an ingredient or the application of an agent in a measured dose. In terms of fumigation, it is the number of ounce-hours (or gram-hours) accumulated during the exposure period.
Dose	The amount of fumigant introduced into the fumigation space. In the case of the metal phosphides, the weight of solid material is not the dose; the dose is the amount of gas generated from the solid, most commonly listed in grams.
Drift	The movement of pesticide dust, spray, or vapor through the air away from the application site.
Easement	In sewer work, the location of a sewer line in backyards, parks, public lands, off-road locations, or other areas which are typically more difficult to access than sewers located beneath street surfaces. Also, the right of utility companies and municipal agencies to access manholes and sewer lines which are located on private property.
Eduction tube	In liquid gas cylinders, a tube that extends to the bottom of the cylinder, enabling liquid instead of gas to be drawn out.
Effluent	The treated wastewater that leaves a sewage treatment plant.
Endangered species	Any species that is in danger of extinction throughout all or a significant portion of its range (i.e., normal area it lives in).
Endangered species bulletin	A part of EPA's Endangered Species Protection Program. Bulletins referenced on the product label that set forth geographically specific pesticide use limitations for the protection of threatened and endangered species and their designated critical habitats.
Environmental Protection Agency (EPA)	The federal agency responsible for registering pesticides and regulating pesticide use in the United States.
EPA registration number	The number assigned to a pesticide by the U.S. EPA. This number must appear on the pesticide label of all registered pesticides.
Eradication	The pest management strategy that attempts to eliminate all members of a pest species from a defined area.
Equilibrium	Even distribution. A fumigant has reached equilibrium when there is an equal concentration of gas throughout a given space. This can be accomplished through the use of "mixing" fans at the time of initial application.
Exclusion	A pest management technique that uses physical or chemical barriers to prevent certain pests from getting into a defined area.
Exfiltration	The leakage of wastewater from a sewer pipe into the ground through joints, cracks, or defects.
Exhaust stack	A duct used to exhaust the fumigant from the treatment area.
Exposure	When an individual comes into contact with a pesticide either through skin, eyes, ingestion, or breathing it (or its vapors) in.
First aid	The immediate assistance provided to someone who has received an exposure to a pesticide. First aid for pesticide exposure usually involves removal of contaminated clothing and washing the affected area of the body to remove as much of the pesticide material as possible. First aid is not a substitute for competent medical treatment.
Fit check	An on-the-spot check to make sure that a fit-tested respirator or self-contained breathing apparatus (SCBA) still fits correctly. A fit check should be done each time the respirator or SCBA is worn.
Fit test	A qualitative test that must be done before you use a respirator or SCBA for the first time and which will determine whether or not the device fits well enough to adequately protect you during use.
Fission	The process by which a bacterial cell reproduces by splitting into two identical cells
Flammability	The ability to support combustion (i.e., the process of burning). A flammable substance easily catches fire.

Flow	The actual amount of water flowing by a particular point over some specified time.
Foaming agent	An adjuvant used to convert a pesticide solution into a thick foam. Used in sewer line root control as a carrier and surface active substance that forms a fast-draining foam to provide maximum contact with the plant surface, to insulate the surface, and reduce rate of evaporation.
Formulation	A mixture of active ingredient combined during manufacture with inert materials. Inert materials are added to improve the mixing and handling qualities of a pesticide.
Fruiting bodies	Special structures produced by fungi that contain and release reproductive spores.
Fumigant	Pesticides in a gaseous state used to penetrate porous surfaces to control pests in containers, enclosed areas, storage facilities, and soil. Fumigants are toxic when absorbed or inhaled.
Fumigation	The process of using a fumigant to control certain pests by exposing them to an atmosphere of toxic gas inside an enclosed area or under tarps.
Fumigation Management Plan	A written plan for a specific fumigation that is prepared before the start of the fumigation.
Fungicide	A pesticide that kills or controls fungi.
Fungus	A multicellular microorganism and common agent of wood decay. The fungus fruiting body (such as a mushroom) consists of filamentous strands called mycelia and reproduces through dispersal of spores (plural: fungi).
Gas / gases	A state of matter consisting of particles that have neither a defined volume nor defined shape. As vapors, gases do not leave residues.
Grade	A measure of relative elevation from one area to another.
Green wood	Freshly sawn or undried wood that still contains tree sap. Wood that has become completely wet after immersion in water would not be considered green, but may be said to be in the "green condition."
Half-loss time (HLT)	The amount of time it takes to lose half the amount of fumigant from an enclosed space as a result of leakage, breakdown or sorption.
Halide leak detectors	A detector for monitoring the presence and approximate concentration of halide gases. Used to measure methyl bromide or sulfuryl fluoride.
Handler	For the scope of this study manual, this is a person who mixes, loads, transfers, applies, or assists with the application of pesticides; who maintains, services, repairs, cleans, or handles equipment used in these activities; who works with unsealed pesticide containers; who adjusts, repairs, or removes treatment site coverings; or who performs other handling activities specified by the product label. It does not include inspection, sampling, or other similar official duties performed by local, state, or federal officials.
Hazard	The amount of danger to people or the environment posed by a pesticide or other toxic material.
Hazardous materials	Materials, including many pesticides, that have been classified by regulatory agencies as being harmful to the environment or to people. Hazardous materials require special handling and must be stored and transported in accordance with regulatory mandates.
Hazardous waste	Hazardous material for which there is no further use. Remains from pesticide spill cleanup are often hazardous wastes. Hazardous wastes can be disposed of only through special hazardous material incineration or by transporting to a Class 1 disposal site.
Heartwood	The wood extending from the pith to the sapwood, the cells of which no longer participate in growth of the tree. Heartwood may contain phenolic compounds, gums, resins, and other materials that usually make it darker and more decay-resistant than sapwood.
Herbicide	A pesticide toxic to plants.
Histoplasmosis	A disease caused by the fungus *Histoplasma capsulatum*. Symptoms of this infection vary greatly, but the disease primarily affects the lungs and occasionally, other organs.
Hyphae	Long, branching filamentous structures of fungi. In most fungi, hyphae are the main mode of vegetative growth, and are collectively called a *mycelium*.
Inert ingredients	All materials in the pesticide formulation other than the active ingredient. Some inert ingredients may be toxic or hazardous to people.
Infiltration	The process by which groundwater enters sewer systems through joints or other defects.

Infiltration/inflow control (i/i)	In general, the process of abating or controlling the introduction of extraneous water in a sewer system. Examples include grouting, re-lining, manhole rehabilitation, etc.
Inhalation	The method of entry of pesticides through the nose or mouth into the lungs.
Interceptor sewer	Typically, a large diameter sewer without service connections which receives wastewater from collector sewers and transports the flows to a wastewater treatment plant.
Integrated Pest Management (IPM)	A pest management program that uses life history information and extensive monitoring to understand a pest and its potential to cause economic damage. Control is achieved, when practical, through multiple approaches including prevention, cultural practices, pesticide applications, exclusion, natural enemies, and host resistance. The goal is to maintain long-term suppression of target pests with minimal impact on non-target organisms and the environment.
Invert	For sewers, this is the lowest point of a pipeline or conduit. The bottom part of a manhole that is rounded to conform to the shape of the sewer line.
Irreversible injury	A health condition caused by certain exposures to toxic pesticides and other hazardous materials from which there is no medical cure or recovery.
Joint	The connection between two adjacent pieces of sewer pipe.
Label	The information, including directions for use, restrictions, requirements and safety procedures, printed on or attached to the pesticide container or wrapper. This information is legally binding.
Labeling	The pesticide container label and all associated materials, including supplemental labels, special registration labels, other items referred to on the label (i.e., applicator's manual), and other manufacturer's information such as brochures and fliers.
Lateral sewer	See building sewer line.
Lignin	The second most abundant constituent of wood cell walls. Lignin gives wood strength and rigidity.
Lineal feet	A measurement of distance, in a straight line, between two adjacent manholes in a sewer system.
Load factor	The amount of fumigant sorbed by the materials being fumigated.
Metabolism	All the chemical processes that occur within a living organism in order to maintain life.
Microorganism	An organism of microscopic size, such as a bacterium, virus, fungus, or alga.
Millions of gallons per day (MGD)	Used to express the design flow capacity or actual flow of a wastewater treatment facility.
Mode of action	The way a pesticide reacts with a pest to eliminate it.
Molecular weight	A measure of the sum of the atomic weight values of the atoms in a molecule.
Molecule	The smallest particle of a substance that retains all the properties of the substance and is composed of one or more atoms.
Mycelium	The vegetative body of a fungus, consisting of a mass of slender filaments called hyphae. (plural: mycelia.)
Non-systemic	A contact pesticide which has a localized pesticidal effect; not transported through the plant or animal tissues.
Non-selective	A pesticide that has an action against many species of pests rather than just a few.
Ocular	Pertaining to the eye—this is one of the routes of entry of pesticides into the body.
Oral	Through the mouth—this is one of the routes of entry of pesticides into the body.
Overflow	An undesirable discharge of sanitary or combined sewer flow into a river, stream, or other surface water.
Particulate	Microscopic particles of solid or liquid matter suspended in the air. NOT the same as a vapor or gas.
Permit conditions	Stricter requirements than those on a pesticide's labeling or in California law and regulation. Permit conditions are issued by the CAC and must be followed.

Parts per million (ppm)	A typical measure of the concentration of a pesticide in another substance. For example, one gallon of active ingredient in 1,000,000 gallons of water represents a 1 ppm concentration.
Personal protective equipment (PPE)	Devices and apparel worn to protect the body from dermal, eye, and inhalation exposure to pesticides or pesticide residues. These include coveralls, eye protection, gloves and boots, respirators, aprons, and hats.
Pesticide	Any substance or mixture of substances intended for preventing, destroying, repelling, or mitigating any insects, rodents, nematodes, fungi, or weeds, or any other forms of life declared to be pests, and any other substance or mixture of substances intended for use as a plant regulator, defoliant, or desiccant. In California, spray adjuvants are also considered pesticides.
Pesticide resistance	The genetically acquired ability of an organism to tolerate the toxic effects of a pesticide.
Photosynthesis	The process by which plants convert sunlight into energy.
Posting	The placing of signs around an area to inform workers and the public that the area is being, or has been, treated with a pesticide.
Post-application summary	This document describes any actions or events that occurred during the fumigation that differed from the FMP, measurements (e.g., humidity) taken to comply with Good Agricultural Practices (GAPs), if not recorded in the FMP, the National Weather Service forecast during application and 48 hours following application, and any incidents or complaints.
Preservative	Any pesticide that, for a reasonable length of time, is effective in preventing the development and activity of wood-rotting fungi, borers of various kinds, and harmful insects that deteriorate wood.
Property operator	Shorthand for "operator of the property" as defined in Title 3, California Code of Regulations. This term means a person who owns the property and/or is legally entitled to possess or use the property through terms of a lease, rental contract, trust, or other management arrangement.
Quarantine fumigation	A fumigation ordered by a governmental agency to protect domestic agriculture and forestry from pests (i.e., insects, fungi, viruses) capable of causing catastrophic damage. Quarantines are particularly used to prevent the entry and establishment of foreign species that have no natural domestic enemies (e.g., predators, parasites or diseases).
Receiving waters	The bodies of water into which wastewater treatment plants or storm sewers discharge.
Regulations	The guidelines or working rules that a regulatory agency uses to carry out and enforce laws.
Residue	Traces of pesticide that remain on treated surfaces.
Respirator	A device that filters out pesticide dusts, mists, and vapors to protect the wearer from respiratory exposure during mixing and loading, application, or while entering treated areas while the airborne concentration of fumigants is above the safe concentration as prescribed by the pesticide label. These may either provide a source of clean air (an air-supplying respirator) or filter out particles and/or vapors from contaminated air (an air-purifying respirator).
Restored pressure fumigation	A method of vacuum chamber fumigation where the pressure is lowered, fumigant introduced and then pressure is restored. This is also called "below NAP" (normal atmospheric pressure).
Restricted materials	California classification for highly hazardous pesticides, including federal restricted use pesticides and certain other pesticides that can be purchased, possessed, and used only by certified applicators.
Restricted materials permit	A permit, issued by the County Agricultural Commissioner, to authorize certified applicators to purchase, possess, and use California restricted materials.
Restricted-use pesticide (RUP)	A pesticide designation by U.S. EPA because it may generally cause, without additional regulatory restrictions, unreasonable adverse effects on humans, domestic or wild animals and/or the environment (including injury to the applicator). A restricted-use pesticide may be used only by a certified applicator or non-certified handlers under the direct supervision of a certified applicator (some product labels may require all handlers to be certified applicators).
Risk	Risk, in terms of pesticide exposure, is a measure of the likelihood that a person will be harmed by the pesticide and its particular use. It is a product of both the pesticide's toxicity and the amount of exposure in terms of volume of pesticide and length of time.
Route of exposure	The way a pesticide gets onto or into the body. The four routes of exposure are dermal (on or through the skin), ocular (on or in the eyes), respiratory (into the lungs), and ingestion (through swallowing).
Safety Data Sheet (SDS)	An information sheet provided by a pesticide manufacturer describing chemical qualities, hazards, safety precautions, and emergency procedures to be followed in case of a spill, fire, or other emergency.

Sanitary sewer	A sewer designed to carry only residential or commercial waste, as opposed to a storm sewer.
Saponify	To convert fat or grease to soap by reacting with an alkali such as sodium hydroxide.
Sapwood	The soft outer layers of recently formed wood between the heartwood and the bark, containing the functioning vascular tissue.
Seal / sealing	To enclose an area so that fumigant gas cannot escape too quickly. A good seal will contain a lethal amount of gas long enough to kill the target pests. Sealing is often done with plastic sheeting, tape and adhesives.
Seasoning	The process of removing moisture from green wood to improve its serviceability. Wood may be either air dried (exposed to ambient air in a yard or shed) or kiln dried (dried in a kiln with the use of artificial heat).
Selective	A pesticide that has a mode of action against only a single or small number of pest species.
Self-contained breathing apparatus (SCBA)	A type of respirator that supplies fresh air from an outside or portable source such as a cylinder under pressure. Air enters a mask that tightly covers the entire face.
Sewer section	The length of sewer pipe connecting two manholes.
Signal word	One of three words (DANGER, WARNING, CAUTION) found on every pesticide label to indicate the relative hazard of the chemical.
Site	The area where pesticides are applied for control of a pest.
Skull and crossbones	The symbol on pesticide labels that are highly poisonous. Always accompanied by the signal word "DANGER" and the word "POISON."
Slime	A combination of fungi, algae, bacteria, and other organisms.
Slope	A surface of which one end or side is at a higher level than another; a rising or falling surface.
Soft rot	A type of decay developing under very wet conditions (as in cooling towers and boat timbers) in the outer wood layers. It is caused by cellulose-destroying microfungi that attack the secondary cell walls but not the intercellular layer.
Solubility	The ability of a substance to dissolve in another substance (e.g., salt in water).
Sorption	A physical and chemical process by which one substance becomes attached to another. The term covers both absorption and adsorption.
Specific gravity	The ratio of the density of any substance to the density of some other substance taken as standard, water being the standard for liquids and solids, and hydrogen or air being the standard for gases.
Spore	A reproductive structure produced by some plants and microorganisms that is resistant to environmental influences.
Spot fumigation	A fumigation technique applied to a restricted or localized space within a larger structure that has no connection to other parts of the structure so that area can be separately sealed and fumigated. Often used in food-processing plants, mills, etc.
Start of the fumigation	The point in time at which the fumigant is first introduced/delivered/dispensed into the air of the treatment area.
Sterilization	A process that kills all forms of life, such as fungi, bacteria, viruses, and including spores of fungi and bacteria that happen to be present on a surface.
Storm sewer	A sewer designed to carry only rainwater, groundwater, or surface water.
Stratify / Stratification	The creation of layers of gas within a confined area most often due to physical characteristics such as temperature, humidity and the relative weights of different gases.
Structural fumigation	The process of sealing a building and applying a fumigant to control pests.
Structural pest control	As defined in California law, it is pest control of household pests (including rodents, vermin, and insects), wood destroying organisms, and other pests which may invade households or other structures, including railroad cars, ships, docks, trucks, airplanes, or the contents thereof. It includes the use of pesticides (including insecticides, rodenticides, and fumigants) to control these pests. Persons performing structural pest control must be licensed by the Structural Pest Control Board,

Supplied-air respirator (SAR)	A device that supplies air from a compressed air tank that is located outside of the fumigation area. SARs permit people to enter oxygen-deficient areas or areas where there are highly toxic pesticide vapors.
Surfactant	An adjuvant used to improve the ability of the pesticide to stick to and be absorbed by the target surface.
Sustained-vacuum fumigation	A method of vacuum chamber fumigation where pressure inside the chamber is reduced, fumigant is introduced, and the lower pressure is maintained throughout the fumigation.
Symptom	Any abnormal condition following a pesticide exposure that can be seen or felt or that can be detected by examination or laboratory tests.
Swale	A dip or sag in a sewer pipe, in which water and debris often collects.
Systemic pesticide	A chemical that is absorbed and translocated within an animal or plant to destroy it. Some systemic pesticides are designed to protect the plant or animal against other pests.
Tape-and-seal	A method of sealing an area for fumigation that does not require a tarpaulin. Potential sites for leaks are sealed with polyethylene sheeting and tape.
Target	Either the pest that is being controlled or surfaces within an area that the pest will contact.
Tarpaulin / tarp	A large vinyl-coated nylon, canvas, or polyester sheet used to seal a structure or other spaces for fumigation. "Tarp" is the more informal and more often used term for "tarpaulin." The most common material is polyethylene and the least common is canvas.
Thermal conductivity analyzers	An instrument designed to measure the concentration of fumigant gases within a chamber or other enclosure during fumigation. Also called "TC analyzers."
Threatened species	A group of organisms likely to become endangered in the foreseeable future.
Threshold limit value (TLV)	The airborne concentration of a pesticide in parts per million (ppm) that produces no adverse effects over time.
Toxic	Acting as, or having the effect of, a poison. A toxic substance is one that can cause harm to organisms or the environment.
Toxicant	A substance that, at a sufficient dose, will cause harm to a living organism.
Toxicity	A measure of the capacity of a pesticide (or other substance) to cause injury. That injury can occur soon after the exposure (acute) or appear later than 24 hours following pesticide exposure (delayed and/or chronic effects).
Translocation	The movement of pesticides from one location to another within the tissues of a plant.
Trade name	A brand name of a pesticide. For example, ProFume ® is a trade name for sulfuryl fluoride. The same active ingredient may be sold under different trade name.
Treatment area	The structure, area, or space which is, or was, enclosed or sealed to contain the fumigant during fumigation and continuing until the commodity or structure is moved or released (control and responsibility passed to the owner of the commodity or structure or another person designated by the owner).
Treatment buffer zone	An area surrounding a treatment area during the treatment period (exposure or holding period) where access is limited. The treatment buffer zone extends from the perimeter of the treatment area to a distance determined by the product label. The treatment buffer zone begins when the fumigant is introduced into the fumigation enclosure and ends when aeration begins.
Trigger levels	The specific air concentration of a fumigant that triggers the requirement for a fumigant handler to use specific respiratory protection in order to continue working in the area being fumigated.
Vacuum chamber	A specially designed structure used to conduct a fumigation where the air can be removed prior to introducing the fumigant.
Vapor	A substance in the gaseous state as distinguished from the liquid or solid state. As a vapor it exists in the air as separate molecules.
Vapor pressure	The pressure exerted by vapor molecules in equilibrium with its condensed phases (solid or liquid) at a given temperature in a closed system. The pressure exerted by a liquid or a solid as it volatilizes (becomes a gas).
Vertebrate	Animals that contain a backbone.
Virus	A very small organism that multiplies inside living cells and is capable of producing disease symptoms in some plants and animals.

Volatile	A substance that vaporizes readily. That means it changes from a liquid (or even a solid) into the gas phase.
Volatility	A measure of how easily a substance vaporizes.
Volatilization	The process or act of vaporizing.
Warning agent	In fumigation, a substance added to the fumigation space to warn or deter people from entering and remaining in the area during the fumigation. The usual substance used is chloropicrin because of its easily recognized smell and sensory irritation, such as watery eyes.
Warning placard / sign	A sign that must be posted at all external entrances and all sides of a structure warning that the structure is being fumigated. Sometimes "sign" and "placard" are used interchangeably.
Water table	The upper level of a groundwater reservoir or aquifer.
Weed	Any plant that interferes with the growing of crops or ornamental plants, endangers livestock, affects the health of people, interferes with the safety or use of roads, utilities, and waterways, or is a visual or physical nuisance.
Wettable powder	A type of pesticide formulation consisting of an active ingredient that will not dissolve in water combined with a mineral clay or other inert ingredients and ground into a fine powder.
White rot	In wood, any decay or rot that attacks both the cellulose and lignin and produces a generally whitish residue, which may be spongy or stringy. Also may occur as pocket rot.

Index

A
Absorption 7
Adsorption 7
Aeration
- and half-loss time 79
- and load factor 78
- Air monitoring 39, 79
- devices used for 41
- precautions for 43

Air movement
- effects on fumigation 30

Aluminum flasks
- proper disposal of 26

Aluminum phosphide 8
Ambient air analyzers 41
Applicator manual
- and labels 21

Area and volume
- calculations 32

B
Bacteria 170
Beetles
- as grain pests 159
- comparison of beetle and moth larvae 162
- wood-boring 153

Boiling point 5
Breathing zone 51
Buffer zones 52
- for methyl bromide 52

Burrow fumigation 77, 95

C
Calculating area and volume 32
Carbon dioxide 12
- use in burrows 95
- uses of 12

Chamber fumigation 83
Chemical reactivity 6
- of methyl bromide 9, 12
- of phosphine 8
- of sulfuryl fluoride 11

CT Concept 32

D
Desorption 7

Detector tubes 42
Diffusion 7
Disposal
- of aluminum flasks 26
- of leftover fumigant 26

E
Endangered Species Bulletins 98

F
First aid
- for fumigant exposure 49

Flammability 6
Fumigants
- advantages and disadvantages 3
- calculating dosage for 32
- concentration requiring leak repair 40
- concentrations requiring respirator use 71
- diffusion of 7
- lethal effects 3
- monitoring for 39
  - devices 41
- physical and chemical characteristics of 4
- storage of 23
- symptoms of exposure to 47
- transporting of 23

Fumigation
- effects of humidity on 31
- effects of site characteristics and environmental conditions 30
- effects of sorption on 7
- methods of 73
- of grain 88
- of boxcar and trucks 112
- of ships 112
- premise inspection for 29
- securing the site 52
- stack 90
- Steps involved 73

Fumigation Management Plan (FMP) 54
- and equipment failure 80

Fungi 169

G
Gas
- and vapor compared 2

Gas cartridge 95
Grain
- fumigation of 88
- pests of 157

Grain bins
- fumigation of 88
- safety 89

Ground seal 104

H

Half-loss time 79

Halide leak detectors 41

Humidity
- effects on fumigation 31

I

In-transit fumigation 113

L

Label 21
- and Applicator Manual 21

Liquefied gases 3

Load factor 78

M

Magnesium phosphide 8
- use in burrows 95

Metam sodium 13
- PPE for 62, 120
- precautions 13
- symptoms of exposure to 49
- use in sewer root control 119
- uses of 13

Methyl bromide 12
- chemical reactivity of 13
- phaseout 12
- PPE for 62
- precautions 12
- symptoms of exposure to 48
- uses of 12

Microorganisms 168

Molecular weight 4

Monitoring 39
- devices used for 41
- efficacy and safety monitoring 40

Moths comparison of beetle and moth larvae 158

N

Notification
- and FMP 43
- for in-transit fumigation 113

P

Personal Protective Equipment (PPE) 59
- comparison of PPE for fumigants and non-fumigants 59
- for metam sodium 120

Pesticide resistance 15

Pest identification
- importance of 14

Pests
- factors of that influence fumigation 14
- termites 147
- wood-boring beetles 153
- stored commodity pests 157
- rodents 165

Phosphine 8
- chemical reactivity 10
- fire and explosion risk 10
- introduction of tablets or pellets 76
- odor 9
- PPE for 61
- precautions when using 10
- residues 10
- symptoms of exposure to 48
- uses of 9

Placards 53

Plastic sheeting 104

Post-Application Summary (PAS) 55

Premise inspection 29

Primary and secondary feeders 157

Q

Quarantine and pre-shipment fumigation 77

Quarantine fumigation 91

R

Remedial Wood Protection 139

Residue
- from phosphine 10

Respirators 63
- and sensory irritation 51
- employer responsibility for 63
- fit test and fit check 68
- medical evaluation for 68

Rodents 165

S

Sealing 38, 89, 104
- and FMP 55
- of grain bins 89

-tape-and-seal 38
Sensory irritation 51
Sewer root control 115
Ships
-fumigation of 111
Solubility 6
Sorption
-effects on fumigation 7, 31
Specific gravity 4
Spot fumigation 74, 107
Stack fumigation 90
Sulfur Dioxide 13
-PPE for 66
-precautions for 14
-symptoms of exposure to 49
-uses of 14
Sulfuryl fluoride 11
-chemical reactivity of 11
-PPE for 62
-precautions for 11
-symptoms of exposure to 49
-uses of 11
Symptoms
-of fumigant exposure 47

T

Tarpaulins (tarps) 38, 74, 104
-and aeration 80
Temperature
-effects on fumigation 31
-effects on grain fumigation 88
Termites 147
-control of with fumigants 151
Thermal conductivity analyzers 42
Toxicity
-risk and exposure 46
Transport fumigation 112

V

Vacuum chamber fumigation 84
Vapor
-and gas compared 2
Vapor pressure 4

Volatile
-defined 4

W

Warning signs 53
-removal of 54
Wine Barrel and Cork Sanitization 135

References

California Department of Pesticide Regulation. (2020). Laws and Regulations Study Guide, 3rd Edition. California Department of Pesticide Regulation.

Department of Labor Logo United States Department of Labor. Respiratory Protection - Respirator Selection - Air-purifying vs. Atmosphere-supplying Respirators | Occupational Safety and Health Administration. (n.d.). Retrieved February 9, 2023, from https://www.osha.gov/etools/respiratory-protection/respirator-selection/air-purifying-atmos-supply

Department of Labor Logo United States Department of Labor. Respiratory Protection - Respirator Selection - Air-purifying vs. Atmosphere-supplying Respirators | Occupational Safety and Health Administration. (n.d.). Retrieved February 9, 2023, from https://www.osha.gov/etools/respiratory-protection/respirator-selection/air-purifying-atmos-supply

Duke, K., & Jessen, E. (1996, March). Sewer Line Chemical Root Control - With Emphasis on Foaming Methods using Metam Sodium and Dichlobenil. Retrieved August 2022, from https://cms.agr.wa.gov/WSDAKentico/Imported/SewerRootCtrlManual.pdf?/SewerRootCtrlManual.pdf

Non-Soil Fumigation – A National Pesticide Applicator Study Manual published by the Pesticide Educational Resources Collaborative (PERC), September 2021

Thomasson, G., Capizzi, J., Dost, F., Morrell, J., & Miller, D. (2015, December). Wood Preservation and Wood Products Treatment. Retrieved August 2022, from https://catalog.extension.oregonstate.edu/sites/catalog/files/project/pdf/em8403_1.pdf

University of California, Statewide Integrated Pest Management Project. (1996). Sewer Line Root Control. University of California.

University of Kentucky - College of Agriculture, Cooperative Extension Service. (n.d.). Sewer Root Control Category 16. Retrieved August 2022, from https://entomology.ca.uky.edu/files/catmanualpdfs/16-manual.pdf

U.S. Department of Agriculture. (n.d.). Fungal decay hazard map - apawood.org. Research in Progress. https://www.apawood.org/data/sites/1/documents/technicalresearch/rip/fplrip-4723-022.pdf

WHITHAUS, S., & BLECKER, L. (2016). The Safe and Effective Use of Pesticides, 3rd Edition. University of California Agriculture and Natural Resources, Publication 3324.